Differentiated Instruction for Science

Instructions and activities for the diverse classroom

Dawn Hudson

The classroom teacher may reproduce materials in this book for classroom use only.
The reproduction of any part for an entire school or school system is strictly prohibited.
No part of this publication may be transmitted, stored, or recorded in any form
without written permission from the publisher.

1 2 3 4 5 6 7 8 9 10
ISBN 0-8251-5896-6
Copyright © 2006
J. Weston Walch, Publisher
P.O. Box 658 • Portland, Maine 04104-0658

Printed in the United States of America

Table of Contents

Introduction .... *iv*
Using the Teacher's Pages .... *vii*
Chart of Differentiated Instruction Techniques .... *viii*
Chart of Multiple Intelligences .... *ix*

Chapter 1: Physical Science: Properties and Changes of Properties in Matter ... 1
1. What's the Matter? 2
2. How Dense Are You? 5
3. The Case of the Mixed-up Powders 9

Chapter 2: Physical Science: Motion and Forces 12
4. Why Should You Wear a Seat Belt? 13
5. Newton's Laws of Motion 16
6. Electromagnetism 19

Chapter 3: Physical Science: Transfer of Energy 22
7. Electric Slide 23
8. Do You Hear What I Hear? 25
9. Convection, Conduction, and Radiation Station 29

Chapter 4: Life Science: Structure and Function in Living Systems ... 32
10. Structure and Function Creatures 33
11. Onions Versus Whitefish 36

Chapter 5: Life Science: Reproduction and Heredity 39
12. Extracting DNA from Cheek Cells 40
13. Protein Manufacturing 44

Chapter 6: Life Science: Regulation and Behavior 47
14. How Much Is That Ficus in the Window? 48
15. Feeling Blue Today? 51

Chapter 7: Life Science: Populations and Ecosystems 56
16. Measuring Up to a Blue Whale 57
17. Don't Get Caught in the Food Web! 60

Chapter 8: Life Science: Diversity and Adaptations of Organisms ... 63
18. Giraffes Can't Jump! 64
19. Extincting Extinctions 67

Chapter 9: Earth and Space Science: Structure of the Earth System ... 70
20. The Water Cycle Song 71
21. Rocks Bingo 73
22. Cloud Cover Song 76

Selected Answers 79
Bibliography 81

Introduction

To meet the needs of all students and design programs that are responsive to the intellectual strengths and personal interests of students, we must explore alternatives to traditional science instruction. We need to examine not only what is taught but how it is taught and how students learn.

Carol Ann Tomlinson in *The Differentiated Classroom: Responding to the Needs of All Learners* encourages educators to look at teaching and learning in a new way. Using the phrase "One size doesn't fit all," she presents, not a recipe for teaching, but a philosophy of educational beliefs:

- Students must be seen as individuals. While students are assigned grade levels by age, they differ in their readiness to learn, their interests, and their style of learning.

- These differences are significant enough to require teachers to make accommodations and differentiate by content, process, and student products. Curriculum tells us what to teach; differentiation gives us strategies to make teaching more successful.

- Students learn best when connections are made between the curriculum, student interests, and the students' previous learning experiences.

- Students should be given the opportunity to work in flexible groups. Different lessons point toward grouping students in different ways: individually, heterogeneously, homogeneously, in a whole group, by student interests, and so forth.

- There should be ongoing assessment—assessment can be used to help plan effective instruction.

To address the diverse ways that students learn and their learning styles, we can look to Howard Gardner's eight intelligences to provide a framework. Gardner's theory of multiple intelligences encourages us to scrutinize our attitudes toward science learning so that each student can learn in a more relaxed environment.

Let's explore what multiple intelligences look like in the science classroom.

Visual/Spatial
Perceives the visual world with accuracy; can transform and visualize three dimensions in a two-dimensional space. Encourage this intelligence by using graphs and making sketches, exploring spatial visualization problems, relating patterns in science to visual and color patterns, using mapping activities, and using manipulatives to connect concrete with abstract.

Verbal/Linguistic
Appreciates and understands the structure, meaning, and function of language. These students can communicate effectively in both written and verbal form. Encourage this intelligence by using class to discuss scientific ideas, using journals to explore scientific ideas using words, making written and oral presentations, and doing research projects.

Logical/Mathematical
Ability to recognize logical or numerical patterns and observe patterns in symbolic form. Enjoys problems requiring the use of deductive or inductive reasoning and is able to follow a chain of reasoning. Activities related to this intelligence include organizing and analyzing data, designing and working with spreadsheets, working on critical-thinking and estimation problems, and making predictions based upon the analysis of numerical data.

Musical/Rhythmic
The ability to produce and/or appreciate rhythm and music. Students may enjoy listening to music, playing an instrument, writing music or lyrics, or moving to the rhythms associated with music. Activities related to this intelligence include using songs to illustrate science skills and/or concepts and connecting rational numbers to musical symbols, frequencies, and other real-world applications.

Bodily/Kinesthetic
The ability to handle one's body with skill and control, such as dancers, sports stars, and craftspeople. Students who excel in this intelligence are often hands-on learners. Activities related to this intelligence include the use of manipulatives, involvement with labs and hands-on activities (weighing, measuring, building), and permitting students to participate in activities that require movement or relate physical movements to scientific concepts.

Interpersonal

The ability to pick up on the feelings of others. Students who excel in this intelligence like to communicate, empathize, and socialize. Activities related to this intelligence include using cooperative-learning groups; brainstorming ideas; employing a creative use of grouping (including heterogeneous, homogeneous, self-directed, and so forth); and using long-range group projects.

Intrapersonal

Understanding and being in touch with one's feelings is at the center of this intelligence. Activities related to this intelligence include encouraging students to be self-reflective and explain their reasoning, using journal questions to support metacognition, and giving students quiet time to work independently.

Naturalist

Naturalist intelligence deals with sensing patterns in and making connections to elements in nature. These students often like to collect, classify, or read about things from nature—rocks, fossils, butterflies, feathers, shells, and the like. Activities related to this intelligence include classifying objects based upon their commonalities, searching for patterns, and using Venn diagrams to help organize data.

The Format of the Book

Primarily this book is arranged based on science subjects being taught throughout the nation. This includes the traditional, separated science subject areas such as life science or biology (including environmental science), physical science (or chemistry and physics), and earth science. This book may also be utilized by teachers teaching integrated science curricula. The arrangement within this book will allow teachers to easily choose activities and labs that not only fit with their standards, but also exemplify differentiated science instruction. Embedded throughout the book are strands based on the National Science Education Standards (NSES) developed by the National Research Council (1995) including Inquiry, Science and Technology; Science in Personal and Social Perspectives; and History and Nature of Science.

Using the Teacher's Pages

Each activity is preceded by a "Teacher's Page" that has valuable information for managing the lesson. In addition, if the teacher changes the lesson or connects it to other lessons, these changes may be noted on the Teacher's Page to be used in future lesson planning. Science journals for both laboratory observations in addition to thinking processes are strongly encouraged.

- **Science Topics:** Most science experiments address more than one science skill or topic. In the real world, science is often an integrated experience, and skills and concepts interrelate and blend. When using these experiments, teachers can use this section to connect the lesson to skills and concepts that are part of their science curriculum.

- **Educational Goals:** A brief listing of the lesson concepts based on standards will be listed.

- **Multiple Intelligences:** Teachers and students alike are encouraged to explore a variety of multiple intelligences. While each experiment does not focus on each of the eight intelligences, activities are open-ended and allow students to use a variety of strategies and intelligences to solve the problems.

- **Materials:** A comprehensive list of materials and supplies needed are listed to aid the teacher in laboratory preparation time and organization.

- **Procedure:** These are not intended to be a scripted, step-by-step plan but rather suggestions to help facilitate students and act as motivation for the experiences. Some lessons suggest specific questions to be asked while others offer suggestions to develop student understanding. These are merely suggestions and should be used only if appropriate to the needs of the class.

- **Differentiation:** Some of the activities may be extended, refined, or differentiated. Suggestions have been made in this section. This would also be an appropriate place for teachers to make notes of their own on ways to enhance the lesson or add additional activities.

- **Assessment:** Multiple suggestions are made in this section. These may include traditional quizzes or tests, completed student projects (forms of formal assessment), observation and questioning (or other forms of informal assessment), and so forth.

Chart of Differentiated Instruction Techniques

Science Content Standards	Activities	Jigsaw	Portfolios/Journals	Tiered Lessons/Stations	Varied Supplementary Materials	Tiered Products	Whole Class Instruction	Group Investigation	Small Group Instruction	Independent Study	Problem-based Learning	Varied Homework	Allowing Choice	Interest Centers
Physical Science: Properties and Changes of Properties in Matter	What's the Matter?		•	•	•	•				•	•	•		
	How Dense Are You?		•	•	•	•	•		•				•	•
	The Case of the Mixed-up Powders	•	•	•	•	•	•	•			•			
Physical Science: Motions and Forces	Why Should You Wear a Seat Belt?		•		•		•		•	•	•		•	
	Newton's Laws of Motion		•	•			•						•	
	Electromagnetism		•		•				•		•			
	Electric Slide		•	•										
Physical Science: Transfer of Energy	Do You Hear What I Hear?		•						•	•				•
	Convection, Conduction, and Radiation Station		•	•	•	•	•		•		•			
Life Science: Structure and Function in Living Systems	Structure and Function Creatures					•	•		•				•	
	Onions Versus Whitefish		•				•			•			•	
Life Science: Reproduction and Heredity	Extracting DNA from Cheek Cells		•			•	•	•	•	•	•		•	
	Protein Manufacturing		•				•		•	•		•		
Life Science: Regulation and Behavior	How Much Is That Ficus in the Window?	•	•	•		•	•		•				•	
	Feeling Blue Today?		•	•			•				•			
Life Science: Populations and Ecosystems	Measuring Up to a Blue Whale	•			•		•	•	•		•			
	Don't Get Caught in the Food Web!		•		•		•			•		•		
Life Science: Diversity and Adaptations of Organisms	Giraffes Can't Jump!			•			•			•	•			
	Extincting Extinctions		•	•		•	•		•	•			•	
Earth and Space Science: Structure of the Earth System	The Water Cycle Song		•			•	•						•	•
	Rocks Bingo			•		•	•		•				•	
	Cloud Cover Song					•	•		•				•	

viii

© 2006 Walch Publishing
Differentiated Instruction for Science

Chart of Multiple Intelligences

Science Content Standards	Activities	Visual/Spatial	Verbal/Linguistic	Logical/Mathematical	Musical/Rhythmic	Bodily/Kinesthetic	Interpersonal	Intrapersonal	Naturalist
Physical Science: Properties and Changes of Properties in Matter	What's the Matter?	●	●	●		●	●	●	
	How Dense Are You?	●	●	●		●	●	●	
	The Case of the Mixed-up Powders	●	●	●			●		
Physical Science: Motion and Forces	Why Should You Wear a Seat Belt?	●	●	●		●	●		
	Newton's Laws of Motion	●	●	●	●	●		●	
	Electromagnetism	●	●	●		●	●	●	
Physical Science: Transfer of Energy	Electric Slide		●	●	●	●	●	●	
	Do You Hear What I Hear?	●	●	●	●	●			
	Convection, Conduction, and Radiation Station	●	●			●	●		
Life Science: Structure and Function in Living Systems	Structure and Function Creatures	●	●	●	●	●	●		●
	Onions Versus Whitefish	●	●	●			●	●	
Life Science: Reproduction and Heredity	Extracting DNA from Cheek Cells	●		●		●	●	●	●
	Protein Manufacturing	●	●	●	●	●	●	●	●
Life Science: Regulation and Behavior	How Much Is That Ficus in the Window?	●	●	●	●	●	●	●	●
	Feeling Blue Today?	●	●	●		●	●	●	
Life Science: Populations and Ecosystems	Measuring Up to a Blue Whale	●	●	●	●	●	●	●	●
	Don't Get Caught in the Food Web!	●	●	●		●	●	●	●
Life Science: Diversity and Adaptations of Organisms	Giraffes Can't Jump!	●	●	●		●	●	●	●
	Extincting Extinctions	●	●	●	●	●	●	●	●
Earth and Space Science: Structure of the Earth System	The Water Cycle Song	●	●	●	●	●	●	●	
	Rocks Bingo	●	●	●	●	●	●	●	●
	Cloud Cover Song	●	●	●	●	●	●	●	●

Differentiated Instruction for Science © 2006 Walch Publishing

Physical Science Properties and Changes of Properties in Matter

Students often find the concept of matter confusing. Furthermore, the fact that *mass* and *weight* are "interchangeable" terms on Earth further complicates understanding.

"What's the Matter?" gives students the opportunity to build models and then use them to answer difficult, application-level questions about solids, liquids, gases, and plasma. It also allows students the opportunity for inquiry-based learning since step-by-step directions are not being provided to the students. Questions about substances that seem to be partially a solid and partially a liquid are also raised.

"How Dense Are You?" measures the density of a student in addition to other smaller items. Comparison is also made to water by using salt water in varying concentrations in solution. Vegetable oil and other oils are also compared to honey, maple syrup, and corn syrup. The formula $D = m/V$ is used to explain the relationship of density to mass and volume.

"The Case of the Mixed-up Powders" twists the traditional unknown white powder lab by having students apply what they know about properties of matter to solve a mystery. Common household compounds are utilized to save the teacher time and money in laboratory preparation.

What's the Matter?

SCIENCE TOPICS

matter, mass, weight, solid, liquid, gas, plasma, inquiry using models

EDUCATIONAL GOALS

Students will

- build three-dimensional models of a solid, a liquid, and a gas.
- explain using their models why each state of matter behaves as it does.
- support their explanations through class discussion.
- make predictions based on their models.

MULTIPLE INTELLIGENCES

visual/spatial, verbal/linguistic, logical/mathematical, bodily/kinesthetic, interpersonal, and intrapersonal

MATERIALS

paper towels, toothpicks, gumdrops or miniature marshmallows (or substitute), salt (optional), microscope (optional)

PROCEDURE

Place students into groups of two to encourage interpersonal communication. Give students a paper towel and a supply of marshmallows or gumdrops and toothpicks. Allow students to design atomic models of a solid, a liquid, and a gas on their own. To encourage student inquiry, tell students that their model of a solid must be able to hold the weight of a large object. The model of a liquid should be somewhere in between the model of the solid and the gas in its durability. In other words, the molecules should be spaced farther apart. The model of a gas should just be able to stand on its own. The "molecules" should be spread apart even farther than in the liquid. To direct students as they work, use Socratic teaching techniques, which means answering any student questions with an additional question and not an answer. This will facilitate the inquiry process and let students correct their own mistakes as they go along. After students have had about 15 minutes of uninterrupted work time, ask the entire class which state of matter takes the shape of its container—a solid, a liquid, or a gas? The students should answer a liquid. Give students a chance to change their models if needed.

Properties and Changes of Properties in Matter
Differentiated Instruction for Science

Depending on the accuracy of the models they have created, students may want to change their original plans.

DIFFERENTIATION

More advanced students may also be given the fourth state of matter (plasma) to work on. This is a more difficult model to create since plasma is a highly charged gas. Before students begin working on the models, they could be shown a solid, a liquid, and a gas and reminded that the solid keeps its shape, the liquid takes the shape of its container, and the gas fills the container completely. Additionally, students can be shown salt crystals under a microscope to give them an idea of what they should use for a crystalline model.

Varied Homework

Students who need a more challenging assignment should be asked to make a mixture of white school glue, water, and borax (sodium borate—available in the laundry powder section of most grocery stores) and describe its characteristics. They should be able to describe this mixture as a solid, a liquid, or a gas and support their reasoning. Students who need a less challenging assignment should be asked to make a mixture of cornstarch and water and describe its characteristics. They should also be able to support whether or not they feel it is a solid, a liquid, or a gas. It is better to give less direction to students about the proportions of each of the ingredients within the mixtures. Encourage inquiry and allow students to vary the ingredients while noting in their journals the amounts of each tried and the result.

ASSESSMENT

Assess the construction and durability of the models created. All three models should be based on a cube or crystalline structure; however, the gumdrops should be tightly packed for the solid and very loosely packed for the gas. The gumdrops should be somewhere in between for the liquid. Remind students that the structures of liquids and gases are not crystalline. In liquids, molecules and atoms move around one another but do not move apart. In gases, molecules move almost independently of one another and are far apart.

Assess students' verbal explanations of why the phases of matter behave the way they do and their level of participation in a class discussion.

Assess students' written answers to the questions asked in the activity.

Properties and Changes of Properties in Matter
Differentiated Instruction for Science

© 2006 Walch Publishing

What's the Matter?

The three most commonly studied states of matter are solids, liquids, and gases. Sometimes it is easier to understand concepts such as matter by building or using a model. A *model* is defined as a representation of an object, which is then used to explain things in a new way.

Directions: Place the paper towel on your desk or lab table so that you and your partner can share the toothpicks and gumdrops or miniature marshmallows that your teacher provides. Using the toothpicks to hold together the gumdrops or marshmallows, build a model or representation of a solid, a liquid, and a gas. Build your model of the solid so that it can support the weight of a heavy object. Your teacher will help you test this in class.

In your science journal, draw the objects you created, being sure to label each one. Then answer the following questions on the lines below or on a separate sheet of paper in your science lab journal.

1. Was your solid able to hold a great amount of weight? Why or why not?

   _____
   _____
   _____

2. What could you do to improve your design of the solid? the liquid? the gas?

   _____
   _____
   _____
   _____

3. Make a prediction based on your models about which state of matter allows the most freedom of movement within the structure.

   _____
   _____
   _____

For homework, your teacher may assign you to make a mixture. Record your own lab procedure in your science lab journal. After the lab, record observations in your journal about whether you feel this mixture is best described as a solid, a liquid, or a gas and why you think this is so.

How Dense Are You?

SCIENCE TOPICS

density, mass, volume, viscosity, displacement, buoyancy

EDUCATIONAL GOALS

Students will

- define *density* using their own terms and vocabulary.
- predict whether objects will float or sink when placed in water.
- predict and then later explain layering when various solutions are poured together and allowed to settle over time.
- calculate the density of various objects in the lab including wooden blocks and/or candle pieces.
- calculate the density of a student volunteer.
- compare viscosity (the force required to make a fluid move) to density (how closely molecules are packed together).
- redefine *density* using their own terms and vocabulary.

MULTIPLE INTELLIGENCES

visual/spatial, verbal/linguistic, logical/mathematical, bodily/kinesthetic, interpersonal, and intrapersonal

MATERIALS

industrial-sized trash can (large enough to immerse a student), plastic or inflatable kids' pool for overflow, scale to measure student weight, balance to measure candles or wood blocks, candle pieces, 2-centimeter wooden blocks of various types, stopwatch, isopropyl (rubbing) alcohol, food coloring, olive oil, vegetable oil, safflower oil, corn oil, baby oil, motor oil (various "weights" including 10W30, 20W50, and so forth), corn syrup, maple syrup, hard-boiled egg, salt, water, honey, two 500 mL beakers, graduated cylinders

PROCEDURE

In order to assess students' prior knowledge, allow students to write the definition of *density* in their own terms. After students have had time to write their definitions, ask students to predict whether a hard-boiled egg will sink or float when placed in water. To encourage all students to answer, have students give thumbs up if they think the egg will float and thumbs down if they think it will sink. Do a class demonstration of density by taking a hard-boiled egg and immersing it in a beaker filled with tap water. Take the egg out of the water and then immerse it into a second beaker that is filled with a saturated salt solution. Students may be surprised at the result. Ask them to write why they think this occurred. A whole-class discussion about density should follow this activating strategy/lesson hook.

An inquiry lesson will follow with pairs of students experimenting with materials prepared by the teacher. Possible experiments by the students are described below:

1. Students can drop a wooden block or candle pieces into various concentrations of salt solutions in graduated cylinders to discover buoyancy.

2. Students can drop a wooden block or candle pieces into various types of oil, including various weights of motor oil to look at varying densities of oil. Similarly, other solutions may be compared such as honey, corn syrup, maple syrup, salt water, and freshwater. A stopwatch may also be used to time how long it takes the candle pieces or block to reach the bottom of the graduated cylinder. Balances may be used to mass the various blocks or candles.

3. Any combination of the various solutions can be mixed so students can note layering within a graduated cylinder. For example, oils, water, honey, corn syrup, maple syrup, isopropyl alcohol, salt water, or other solutions brought from home could be compared. The balance and stopwatch may also be utilized.

4. For homework, students can finish writing about their lab observations and reflect on the various conditions that affect how objects float or sink within various solutions. Students should be prepared to discuss their findings in front of the class.

FOLLOW-UP ACTIVITY: MEASURING THE DENSITY OF A STUDENT

Introduce the mathematical formula for density in which density is equal to mass divided by volume. Ask students to write in their science journals methods they think they could use to find the density of their own bodies. After all predictions are written, choose several students to share their ideas with the class. Place an unused industrial-sized trash can into the center of a plastic or inflatable kids' pool. Fill the trash can with water until it cannot be filled any higher. Find a student who is willing to be completely submerged into the water. Find out the mass of the student. If the scale measures in pounds, be sure to convert to kilograms by multiplying by 2.2 lbs/kg. Measure the volume of water displaced into the pool by filling buckets and allowing students to assist in measurement. Mathematically calculate the density of the student using the volume of displaced water and the mass of the student. Take proper safety precautions if performing this experiment.

DIFFERENTIATION

More advanced students may be able to calculate some of the densities of the various wood blocks and the candles used in the lab. Be sure to require that students record their thought processes on the activity sheet or in their science journals. The entire lab devotes itself to differentiation through student choice, tiered products, variation of partner work and whole-group instruction, and interest centers. As students begin to grasp the concept of density, they may choose to do more than one experiment since the materials are so readily available and since they will have a sense of accomplishment as they begin to discover meaning for themselves.

ASSESSMENT

Assess student definitions of density, reflections, and calculations based on the lab activity, student ideas for measuring a student's density, and any calculations made using data gathered in class.

How Dense Are You?

Directions: Answer the following on the lines below or on a separate sheet of paper in your science lab journal.

1. Write your definition of the word *density*.

   _____
   _____

2. When a hard-boiled egg is placed into a beaker of water, do you think it will sink or float? Explain.

   _____
   _____

3. If that same hard-boiled egg is then placed into a container of salt water, do you think it will sink or float? Explain.

   _____
   _____

Design an experiment to be tested in class that uses the supplies provided by your teacher. Follow your teacher's guidelines. The purpose of the experiment you design is to help you draw conclusions about the properties of matter including density. Record all your calculations and observations in your science lab journal. After you have finished your experiment, use what you have learned to answer the following on the lines below or in your science lab journal.

4. Redefine *density* in your own terms. Did your definition change from the first time you attempted to describe it? In what way(s) has it changed and why? (Be prepared to defend your new definition of *density* in front of the class.)

   _____
   _____
   _____

5. Using the knowledge you have gained and the data you have collected during your experiment, devise a way to measure the density of your own body. Explain your reasoning.

   _____
   _____
   _____

The Case of the Mixed-up Powders

SCIENCE TOPICS

mixtures, compounds, inquiry, science-process skills (nature of science), measurement, properties of compounds, classification

EDUCATIONAL GOALS

Students will

- observe and measure characteristic physical and chemical properties of materials such as solubility, flammability, and reaction with various solutions by directly observing with a hand lens.

- distinguish and separate one substance from another (classification) using the characteristic properties of materials.

- design their own experiment in which to test various unknown white powders in order to identify them.

- design a method of organizing data into a chart or graph for comparison in order to draw conclusions from their experiment.

MULTIPLE INTELLIGENCES

logical/mathematical, verbal/linguistic, visual/spatial, and interpersonal

MATERIALS

flour, sugar, salt, baking soda, baby powder, and plaster of paris separated into plastic bags for student groups; vinegar; water; beakers; hand lens; candle; aluminum foil "boat"; tongs or clothespin

SAFETY NOTE

Students should be observed carefully and use caution if they make aluminum foil boats and hold them over a candle flame using tongs or a clothespin. Teachers may choose to do flammability tests as a demonstration instead. Goggles should be worn to prevent vinegar from entering the eyes. Students should not use the sense of taste to identify materials.

DIFFERENTIATION

Since students are designing their own investigation (problem-based learning) to identify the unknown powders, they will have tiered products because of the tiered lesson. The lesson will quickly shift from a whole-class discussion to a group investigation with three or four students per group. The lesson will then shift back to a whole-class discussion in order to assist students in drawing conclusions. Using task cards, the teacher can give less advanced students choices for the powders that are mixed up; other students may be given more choices than actual powders. Students who are more advanced may not be given any choices at all, but they may be asked to identify the powders on their own through experimentation and research.

ASSESSMENT

Informal assessment will occur as the teacher observes students during the design and measurement process with the unknown powders. Teachers can formally assess students' products such as graphs and charts of data, students' conclusions drawn, and whether or not the scientific method was employed throughout the process. If the assignment was differentiated by tiered products, some students may be able to not only list the common names of the powders, but also give the chemical names and/or structures. For example, salt would be sodium chloride or NaCl.

The six powders may be separated by various methods not limited to vision. Salt and sugar grains may be observed with a hand lens; baking soda will bubble due to a chemical reaction with vinegar; salt and sugar will dissolve completely in water in most cases; plaster of paris will harden when water is added; flour and baby powder will make water cloudy; baby powder will smell different from the other powders; flour and baby powder will feel smooth to the touch, whereas salt and sugar will feel grainy due to their crystalline structure; when heated, sugar will begin to melt and flour will burn.

The Case of the Mixed-up Powders

Michelle and Russ are going camping with their baby, Grace. To save space, they decided to empty the contents of bulky containers into smaller plastic bags. The only problem was they forgot to label the bags, and now all the bags are mixed up. Help them separate the bags by using your four senses (excluding taste) so that they do not eat something they should not, or use something in the wrong manner. Remember, some of the materials may not be safe to eat, so the sense of taste should not be used when identifying materials! In this particular case, the senses of smell and touch are appropriate, but not in all laboratory situations.

Directions: Using the scientific method, design ways to identify the six unknown powders given to you by your teacher. Use the materials and guidelines provided by your teacher. It may take some research to find ways to identify each of the substances. In your science lab journal, describe your methods and then record the data you collect throughout your experiments. Be sure to organize your data so that it is reproducible and testable by others.

In the space below, or on a separate sheet of paper in your science lab journal, design a chart, a graph, or another way to clearly display the data you collected.

Do you think you were successful in the identification process? Why or why not?

_____
_____
_____
_____

Properties and Changes of Properties in Matter

Physical Science Motion and Forces

When a force (a push or a pull) is applied to an object, the object changes its motion. Gravity is a universal force that is related to the size of the masses that are being separated. Friction is a force that also acts upon objects. If a force is not acting on an object, the object will continue to move at a constant speed and in a line without stopping. The weight of an object is related to the object's mass. The mass of the object is the amount of matter in the object. Objects that have more mass also weigh more than objects that have less mass. Mass does not depend on gravity, whereas weight does. This is a large misconception among science students! Another misconception is about acceleration. In everyday (nonscientific) terms, acceleration means to speed up. The scientific definition of acceleration includes speeding up, slowing down (negative acceleration), or changing direction. Inertia is related to how difficult it is to change an object's motion. Inertia is also attributed to an object's mass.

The activities in this section will explore Newton's laws of motion.

Newton's First Law: An object at rest stays at rest unless an unbalanced force is acting upon it. If an object is already in motion, it continues to move indefinitely in a straight line unless another force acts upon it.

Newton's Second Law: In mathematical terms, force is equal to the mass times the acceleration. In other words, if an unbalanced force acts upon an object, the object will accelerate in the direction of the applied force. Force is measured in newtons (N). A newton is equivalent to the amount of force it takes to accelerate 1 kilogram of matter 1 meter per second squared.

Newton's Third Law: Forces always act in equal but opposite pairs. The most common way to state this law is that for each and every action force, there is an equal but opposite reaction force.

"Why Should You Wear a Seat Belt?" is an application-level overview of Newton's three laws of motion. Students design their own experiment and then measure how well seat belts hold precious "cargo," which they define. The lab also enables teachers to begin to address misconceptions about inertia, acceleration, and mass versus weight.

"Newton's Laws of Motion" provides a method for teachers to clarify the differences between Newton's three laws through various memory techniques including graphic organizers, foldable notes, and mnemonics. Students will also add their own creative twist by creating songs or poems, or by writing and performing a play, and then presenting to the class. As a final product, students will do a self-reflection for a portion of their grade.

"Electromagnetism" is an inquiry activity in which students attempt to increase the strength of an electromagnet through scientific testing and exploration. Students carefully note the experiment in their science journals by recording data and drawing conclusions.

Why Should You Wear a Seat Belt?

SCIENCE TOPICS

Newton's three laws of motion, inertia, force, mass, acceleration, newton, friction, gravity, weight

EDUCATIONAL GOALS

Students will

- relate Newton's three laws of motion to wearing a seat belt in a vehicle for safety.

- test and gather data about the "seat belt" by examining a clay ball, a raw egg, or another student-chosen "passenger" for their vehicle.

- apply the equation $F = ma$ (where F = force, m = mass, and a = acceleration) to the lab they devise.

- differentiate between mass and weight, inertia, and acceleration as defined in scientific terms.

MULTIPLE INTELLIGENCES

logical/mathematical, verbal/linguistic, bodily/kinesthetic, visual/spatial, and interpersonal

MATERIALS

Materials will vary as determined by the students but could include rubber bands ("seat belts"); eggs, clay balls, or items that might be damaged upon impact ("passengers"); toy cars, trucks, or materials to design a car; balance; metersticks; and stopwatches.

DIFFERENTIATION

Students may be designing their own cars, then deciding both the passenger of the vehicle and the seat belt used. Students will also gather data in a manner they choose and record it in their science journals. The activity is tiered due to the flexibility and choice students have within its design. The teacher will provide a variety of support materials complemented by items the students may bring

from home. Students will work with partners throughout this problem-based learning lab. Because the lab is tiered, the homework will be tiered as well when students experiment with Newton's laws and inertia in relation to a real-life safety example.

ASSESSMENT

Students will be recording their observations in their science journals for review. Students should have data charts to interpret and draw conclusions for Newton's laws. Newton's second law of motion should be discussed as it relates to seat-belt use.

FOLLOW-UP

Teachers should discuss student misconceptions about Newton's laws, acceleration, inertia, and weight versus mass. Holding a discussion with the whole group while students reference their own definitions in their science journals would be helpful. Acceleration is commonly known as "speeding up," but it also can mean slowing down (negative acceleration or deceleration) or a change in direction. Inertia is dependent on the mass of an object. Inertia should be easily observed in this lab while students quickly stop the vehicles and note what happens to the "passengers" with and without seat-belt use. If students do not understand this concept, the teacher can give another example. Ask students the following: If you are stopped at a red light and see two vehicles quickly approaching the back of your car, would you rather the car be a two-seater convertible hitting your car or an 18-wheeler? This example relating inertia to mass will make sense to the students. Lastly, the differences between mass and weight should be noted. The mass of an object is the amount of matter the object contains and does not depend on gravity. Weight, on the other hand, depends on the pull of gravity. If a student weighs 100 pounds or 45 kilograms on the earth, and they were to visit the moon, they would weigh only 17 pounds or 7 kilograms. (To find your weight on other planets or even on a star, visit www.exploratorium.edu/ronh/weight.) Does this mean that students' clothes would fall off them in a pile when they landed on the moon due to the loss of weight? Most students would agree that this would not happen. Their mass did not change; only their weight changed. This is due to a smaller object (the moon) pulling on their mass with less force than the larger earth.

Why Should You Wear a Seat Belt?

One of the first things consumers want to know before they buy a car is its crash test (safety) data. Many of the new cars today have bumpers that can be hit at a low rate of speed without damage to the passenger, driver, or car. Cars are also equipped with antilock brakes, airbags (both front and side), and other safety features. Seat belts keep passengers safe by locking in place so passengers are not ejected from a moving vehicle if an accident occurs.

Understanding all the forces involved around a moving (or quickly stopping) car require studying Newton's laws of motion. This lab will allow you to experiment with forces, acceleration, inertia, mass, weight, and gravity.

Directions: Record definitions (in your own words) for the terms below.

force: _____

acceleration: _____

inertia: _____

mass: _____

weight: _____

gravity: _____

Now work with a partner to design (or attain) a vehicle to test for safety. You an your partner should agree on a "passenger" for your vehicle such as a ball of clay, raw egg, or a substitute of your choice. You will be designing a "seatbelt" to hol your "passenger" in place during a crash test. Be sure to record all data from you experiment in your science lab journal. Suggested instruments to gather data include stopwatches, metersticks, and/or a balance. Remember Newton's second law of gravity: $F = ma$ where F = force, m = mass, and a = acceleration. This law should be explored during your experiment. You may decide to vary first one variable and then another during your experiment until you fully understand how Newton's laws of motion work.

Newton's Laws of Motion

SCIENCE TOPICS

Newton's three laws of motion, inertia, force, mass, acceleration, newton, friction, gravity, weight

EDUCATIONAL GOALS

Students will

- list Newton's three laws of motion and their definitions in a creative manner, focusing primarily on one of the laws in order to explain it to fellow classmates.

- write a skit, a poem, a dance, or sing a song to verify their understanding of Newton's laws.

- design a graphic organizer that relates the terminology of Newton's laws, such as inertia, gravity, mass, weight, force, and acceleration, to other students.

MULTIPLE INTELLIGENCES

logical/mathematical, verbal/linguistic, bodily/kinesthetic, musical/rhythmic, visual/spatial, and intrapersonal

MATERIALS

construction paper, glue, scissors, paper, markers, crayons

PROCEDURE

One student will research Newton's first law of motion well enough to explain it to the other two members of his or her group. The other two members will each take one of the other two laws and do the same. Once all members have had an opportunity to share, they will move to the second part of the project, which is to create a skit, a song, a poem, or a dance to demonstrate understanding of Newton's three laws of motion. Lastly, each individual within the group will design his or her own graphic organizer showing a clear understanding of the relationships of terminology associated with Newton's laws of motion.

DIFFERENTIATION

Students will work in jigsaw groups for this activity. Each student's contribution is important for full understanding of the concept. Other forms of differentiation include varying between whole-class discussion and small-group instruction and then encouraging independent study during the self-reflection. Student choice will boost student morale and increase success in this lab. Tiered products will be the ultimate goal as a result of choices students have in the activity.

ASSESSMENT

Students will be assessed both formally and informally. Informally, students should be assessed by their involvement in educating their fellow classmates about one of Newton's laws. Formally, the skits, plays, and mnemonic devices students create should be assessed for both creativity and accuracy. Students will also receive a grade on the number of connections made through their graphic organizer on terms related to the laws of motion. The last assessment will be a self-assessment in which students will be asked to reflect on their work.

Newton's Laws of Motion

Sir Isaac Newton has been recognized as one of the most outstanding scientists of all time. He formulated three theories about motion that have been tested, tried over and over again, and have withstood the test of time. Today, we call these theories Newton's laws of motion since they are always true.

Directions: Get into groups of three according to your teacher's directions. Research one of Newton's laws of motion, and write the definition on the lines below. Each member of the group should research a different law of motion.

_____

_____

_____

Explain Newton's law well enough so that your partners can understand how all three laws work together to define motion.

Together with your group members, write and perform a skit, write and sing a song, write and read a poem, or choreograph and perform a dance showing you understand Newton's three laws of motion.

On a separate sheet of paper, design a graphic organizer that further proves you understand the finer details of the three laws of motion in addition to other related concepts such as inertia, force, mass, acceleration, weight, friction, and gravity.

In your science lab journal, reflect on your work and describe what you do and do not understand about Newton's laws. Also describe where you feel you did well on the project as well as where you felt you could have done better. Questions to ask yourself about your experiment include the following: Is there a better way to do this experiment? Is there another explanation to support the data gathered? Do I need more evidence to support my conclusions? How reliable/testable is my data?

Electromagnetism

SCIENCE TOPICS

electricity, magnetism, electromagnetism

EDUCATIONAL GOALS

Students will understand

- moving electric charges produce magnetic forces.
- moving magnets produce electric forces.
- the differences between electromagnets and natural magnets.
- the strength of an electromagnet can be changed by passing additional current through a coil of wire or by increasing the number of coils.

MULTIPLE INTELLIGENCES

logical/mathematical, verbal/linguistic, bodily/kinesthetic, visual/spatial, interpersonal, and intrapersonal

MATERIALS

batteries of various voltages, insulated copper wire, iron nails, compass, iron filings, natural magnets, paper clips

PROCEDURE

Students will be working with a natural magnet and then building an electromagnet out of copper wire, a nail, and a battery. Students should answer the questions and record their observations in their science lab journal.

Students should realize they cannot turn off the power of a natural magnet. If a natural magnet is hit enough times with a hammer, however, it may lose some or all of its magnetic ability due to the realignment of the poles. Also, students can turn other metals into magnets in various ways. For instance, if they wanted to magnetize the end of a screwdriver, they would rub a strong magnet in one direction only on the tip of the screwdriver. After a couple of minutes, the end of the screwdriver would be magnetized.

SAFETY NOTE

While building electromagnets, students should use caution when removing plastic coating from copper wire and when handling wire attached to battery. Wire can become hot.

DIFFERENTIATION

Differentiation for this experiment occurs mainly through the tiered lesson and problem-based approach for inquiry learning. Students will be working in pairs with lab partners and will be given a problem to solve. Tiered products will be a natural result of having a tiered lesson. Students will reflect on their experiences in their science lab journals.

ASSESSMENT

Students will be assessed informally as the teacher monitors the progress of students as they begin thinking through the process of solving the problem. Formal assessment will come from their graded science lab journals, in which students will record their trials and errors along with thought processes. The students will also need to reflect on their work.

Electromagnetism

Directions: Using the materials provided by your teacher, follow the directions and answer the questions below. Write your answers and any other observations in your science lab journal.

You will be working with two different types of magnets—a natural magnet and a human-made magnet. You are probably already familiar with a natural magnet. This is the type of magnet you use to hold papers on your refrigerator or on your metal locker. Human-made magnets are known as electromagnets. Electromagnets are unique magnets that can be turned on or off. This ability makes them very useful in our society for such things as picking up scrap metal, loading large trucks onto barges, and so forth.

1. How many paper clips can you pick up with the natural magnet?

2. Could you turn off the magnet's power to pick up or stick to certain metallic objects? How?

3. Can you increase the number of paper clips the natural magnet is able to pick up? Why or why not?

4. Can you make a nonmagnetized object become magnetized? Try it for yourself. Were you successful?

5. Build an electromagnet using copper wire, one or more batteries, and a nail. Wrap the wire tightly around the nail, leaving several inches of loose wire at each end. Do not let the wire overlap as you coil it around the nail. Remove about 1 inch of the plastic coating from each end of the wire. Use tape to attach one end of the wire to one end of a battery. Attach the other end of the wire to the other end of the battery. *Safety note:* Use caution when removing plastic coating from copper wire and when handling wire attached to battery. Wire can become hot. What happens when you attempt to pick up as many paper clips as you can with the nail? Be sure to write your lab procedure in your science lab journal.

6. How does the number of paper clips picked up by this type of magnet compare to the number picked up by the natural magnet?

7. Can you increase the number of paper clips picked up by the electromagnet? How?

8. Place a compass near both the electromagnet and the natural magnet. Record what you observe. Why does this occur?

Physical Science
Transfer of Energy

Energy is transferred in many ways. It is also associated with electricity, heat, light, mechanics, chemical reactions, and sound. The next three activities will take some of these properties and show how energy is transferred.

In "Electric Slide," students will use a popular song and dance to help them memorize Ohm's law. They will apply mathematical principles to science in order to understand voltage, amperage, and current in an electrical system.

In "Do You Hear What I Hear?" sound is discussed in light of the properties of energy it possesses.

"Convection, Conduction, and Radiation Station" places students into the role of a news reporter for WHOT television. In this scenario, students will explore transference of heat between solids, liquids, and gases.

Electric Slide

SCIENCE TOPICS
electricity, Ohm's law, current, resistance, voltage

EDUCATIONAL GOALS
Students will
- dance the Electric Slide while chanting Ohm's law.
- apply Ohm's law to sets of problems.

MULTIPLE INTELLIGENCES
verbal/linguistic, logical/mathematical, musical/rhythmic, bodily/kinesthetic, interpersonal, and intrapersonal

MATERIALS
copy of a song to do the Electric Slide dance to, dance directions printed from the Internet (if needed), and CD or MP3 player for song

PROCEDURE
Clear space in your classroom or go outside or to an open space for this activity. Have students line up in rows behind the person who will be teaching them the dance steps to the Electric Slide. Hand students copies of the student page with the words that will be repeated as students do the Electric Slide. Once students have mastered the dance steps, the words to Ohm's law may be added. Once most students have mastered them together, allow them to work independently on the problems using Ohm's law.

DIFFERENTIATION
Students may work in groups instead of independently. Students may use another form of line dancing with a different song, but the same words to practice this concept. If students are having difficulty using the math formula and converting it from one form to another, you may wish to assist particular students by giving them all three forms of the formula. For example, $V = IR$ will be given to most students, and they will need to change the formula in order to solve for the variables other than V. Less advanced students may be given the other two formulas derived from $V = IR$, $I = V/R$, and $R = V/I$.

ASSESSMENT
1. Assess class participation for learning the Electric Slide and the chant.
2. Review answers to questions on Ohm's law.
3. Review journal reflection from the lesson.

Transfer of Energy
Differentiated Instruction for Science

Electric Slide

Electricity has become so common in our society that it is hard to imagine living without it. Many times at dinner, it is nice to be able to dim the lights to create a softer light. How do dimmer switches work? They rely on a mathematical formula called Ohm's law that relates three principles of electricity: voltage, current, and resistance. Voltage is measured in volts, while current is measured in amps (short for amperes). Resistance is measured in ohms. Ohms are named after Georg Ohm, who did experiments with electricity. Specifically, the formula for Ohm's law can be written $V = IR$, where V is the voltage, I stands for current, and R stands for resistance.

Directions: You will have an opportunity to do the Electric Slide dance today! After your teacher decides you have mastered the steps to the Electric Slide, you will begin to do the Ohm's law chant. Every time the words to the song are "it's electric!" you will instead shout "it's Ohm's law!" Other than those moments in the song, you will continually repeat by singing along with the tune that $V = IR$. For a challenge, if you can manage to do the dance steps and not forget part of the chant, too, you can add the other mathematical derivatives of Ohm's law.

In your science journal, answer the following questions using $V = IR$ or a derivative.

1. If the voltage in a circuit is 120 volts (which is standard in the United States), and the current is 20 amps, what is the resistance of the lightbulb?

2. If, instead, you are plugging in a dryer that has a voltage of 240 volts and the current remains the same at 20 amps, what is the resistance of the dryer?

3. What is the resistance of a heater connected to a 100-volt outlet if the current is 50 A?

4. What is the voltage of a battery that is connected in a circuit in which the bulb has a resistance of 3 ohms and there is a current of 3 A?

Answer the following journal reflection questions.

5. How does doing the Electric Slide and chanting help you to better remember Ohm's law?

6. Applying Ohm's law in your own words, what happens in general as the voltage in a circuit increases?

7. What happens when resistance is increased? Explain your reasoning.

Do You Hear What I Hear?

SCIENCE TOPICS

energy, sound, waves, vibration, amplitude, frequency, pitch, compression, wavelength, rarefaction

EDUCATIONAL GOALS

Students will

- relate sound energy to frequency, pitch, amplitude, wavelength, compression, and rarefaction.

- know sound does not occur without vibrations moving through a solid, a liquid, or a gas.

- understand that waves of any type (sound, water, light) are a form of energy and that energy can be transferred from one form to another.

MULTIPLE INTELLIGENCES

logical/mathematical, verbal/linguistic, bodily/kinesthetic, musical/rhythmic, visual/spatial, and interpersonal

MATERIALS

water, tuning forks (or eating utensil fork may be substituted), CD of recorded sounds, CD player, beaker, Slinky, small pencil boxes or boxes of similar size, rubber bands of various lengths and thicknesses, shoe-box lids of various sizes, Doppler ball or similar

PROCEDURE

- **Station 1: Tuning forks.** Students will hit the tuning fork and place it into a dish of water. (The water should spray out of the bowl and make waves showing the vibrational energy of the tuning fork). Students should record observations in their science lab journals. Students will also put a vibrating tuning fork just behind their ear on the temporal bone. Students should be able to hear the sound vibrating through their heads. This is a good segue to reminding students the reason they sound "funny" when they hear their recorded voice is because they are not hearing it through their skull bones first.

- **Station 2: Listening station.** Make a CD of various sounds to be identified, such as birds, frogs, vacuum cleaner, coughing, train whistle, siren, thunder, rain falling on leaves, various musical instruments, and so forth. You could also have students listen to these sounds on the computer. Students should attempt to guess the sound they hear and record it in their science lab journals. Students should hypothesize how they are able to hear these sounds. If sounds are vibrations, how are sounds able to be heard in the ear? (The eardrum also vibrates, which then sends vibrations into the middle and inner ear to be carried via the auditory nerve to the temporal lobe of the brain for interpretation.)

- **Station 3: Slinky for demonstration of wave types.** Students should move a Slinky in two directions (similar to the motions of playing an accordion) to show how vibrations (energy) in the form of compressions and rarefactions move the energy along. It is important to note misconceptions that arise from this. The particles themselves do not move, but the energy moves through them and along the line.

- **Station 4: Rubber-band music boxes.** Students will build their own "guitars" by stretching rubber bands of various thicknesses and lengths around shoe-box lids. They will discover what thicker rubber bands sound like versus thinner and what shorter rubber bands sound like versus longer. Students need to note their observations in their science lab notebooks. What happens when a guitar string is tightened? The term *pitch* refers to how high or how low a sound is.

- **Station 5: Doppler ball.** Students will throw a Doppler ball and describe the sounds they note while doing so. If a teacher does not have a Doppler ball, one can be made by going to an electronics store and purchasing a battery-operated buzzer. A foam ball from a toy store can be cut large enough to insert the buzzer. Students need to note in their science lab journals what happens when the ball is thrown quickly, slowly, or up in a tall arch.

- **Station 6: Musical instruments.** If possible, band students may be able to bring their instruments into class. The longer the tubing on an instrument, the lower the pitch. For example, a tuba has a lot of tubing as compared to a piccolo. Even within the same instrument, the length of the tubing makes a difference in pitch. Note that as more holes are covered in the flute or clarinet, the pitch lowers because the length of the tube is virtually longer. If musical instruments cannot be brought in, the teacher may wish to purchase small recorders, cymbals, drums, or other "horns" found at discount stores.

- **Station 7: Student choice.** Students research various aspects of sound including decibel level, wavelength, amplitude, frequency, pitch, and so forth to find what each means. Students may design a PowerPoint presentation about what they learned, or they may wish to create a poster, a skit, or a board game.

DIFFERENTIATION

Students complete this activity in a series of lab stations with task cards. The stations allow groups of students to work on different tasks simultaneously. The stations invite students to work on different tasks and encourage flexible grouping. One of the stations allows student interest choices. Students create journal entries after completing tasks.

ASSESSMENT

Students will record data in their science lab journals from various stations as they move from one to the next within a group of three students. Teachers need to make and grade the questions appearing on task cards for each of the seven lab stations.

Do You Hear What I Hear?

This lab will let you explore various lab stations about sound and how sound travels. Record all the data you collect and your observations in your science lab journal.

Directions: Follow the task cards for the lab stations as directed by your teacher. The lab stations may include the following:

- Tuning forks
- Listening station of various sounds
- Slinky for demonstration of wave types
- Rubber-band music boxes
- Doppler ball
- Musical instruments
- Student choice

On Your Own: GPS (Global Positioning System) relies on sound waves to locate objects on Earth by using a satellite. The beam is sent by the GPS, which is picked up by the satellite, reflected, and bounced back with coordinates.

What are some examples of how GPS is changing the world we live in today? Write your answer on the lines below or in your science lab journal. Then present your findings to the class in a form you choose such as a PowerPoint presentation, a poster, or a skit.

_____
_____
_____
_____
_____
_____
_____
_____
_____
_____

Convection, Conduction, and Radiation Station

SCIENCE TOPICS

convection, conduction, radiation, boiling point, freezing point, Fahrenheit, Celsius

EDUCATIONAL GOALS

Students will understand that

- heat moves from a warmer object to a cooler object until both reach the same temperature.
- the sun is a source of light energy in the forms of radiation, visible light, infrared, and UV light.
- when something is heated, the molecules move farther apart (in a flexible container) and speed up.
- when something is cooled, the molecules move closer together (in a flexible container) and slow down.
- metals allow heat to flow freely, whereas glass and plastic hardly allow heat to flow at all.

MULTIPLE INTELLIGENCES

verbal/linguistic, bodily/kinesthetic, visual/spatial, and interpersonal

MATERIALS

beaker or clear pot; water; food coloring; extra large hair or jewelry beads (so when they are placed into boiling water, students in the back of the room will also be able to see them moving); plastic and metal spoons; UV beads that change color in UV light, available from most science suppliers (or items from Del Sol, which also change color with UV light); black light (UV light); night-light with spinning lamp (caused from convection currents) or similar spinning apparatus caused from burning four or more candles on the base; and glitter lamp or lava lamp

SAFETY NOTE

Students should wear goggles to prevent hot water from entering their eyes.

PROCEDURE

Students will role-play news broadcasters and conduct demonstrations on the "news" that show their understanding of convection, conduction, and radiation. Students will write their own script in their science journals, being sure to define and give demonstrations of all these three forms of heat energy. Students will perform for the class the skit they have written. The oral presentation will include the science demonstrations that students have devised about convection, conduction, and radiation.

DIFFERENTIATION

Although students have flexibility, some of the various demonstrations that they may design might be as follows: boiling water with hair beads to actually see the convection currents in the water; adding a drop of food coloring to hot and then cold water to see which dissipates the fastest; and talking about the Celsius scale versus the Fahrenheit scale, where boiling points are 100 degrees and 212 degrees respectively. Freezing points on those scales are 0 degrees and 32 degrees respectively. Body temperature on the Celsius scale is 37 degrees, whereas in Fahrenheit it is 98.6 degrees.

ASSESSMENT

Student presentations will be graded for content. Students should understand that convection is heat moving through either a gas or a liquid, whereas conduction is heat moving through a solid. If students are confused about those terms since they are so similar, remind them that the word *conduction* has the word *duct* in the middle. Then you can ask them if a "duct" is a solid, a liquid, or a gas. The answer is a solid, which is what "conduction" is all about.

Convection, Conduction, and Radiation Station

You are the six o'clock news reporter for a major television station, WHOT, and are about to break an interesting story to your viewers. You have been "teasing" the viewers about an interesting form of energy called heat energy and how it travels or is converted from one form into another.

Directions: Design a news show and present it to your "audience" (the class) to inform them about heat energy. It is up to you to write your script. Your teacher will give you materials that you can use to demonstrate the concepts you are explaining. Include definitions and terms with your news show so the viewers will understand these difficult concepts. Concepts you might address include convection, conduction, radiation, heat, ultraviolet (UV) light, infrared (IR) light, freezing point, boiling point, Celsius, and Fahrenheit. Be sure to include details about each.

After you have presented your news story to the class, answer the following questions on the lines below or in your science lab journal.

1. What term is used to mean heat is leaving an object that is a gas?
   _____

2. What term is used to mean heat is leaving an object that is a liquid?
   _____

3. What term is used to mean heat is leaving an object that is a solid?
   _____

4. How can you remember the difference between convection and conduction, since the terms are so similar?
   _____
   _____
   _____
   _____
   _____
   _____

Life Science
Structure and Function in Living Systems

Understanding the structure and function of various organs and organ systems is vital in understanding living organisms. It also helps scientists begin to ask valid questions about the functions of some structures. For example, what is the purpose of having wisdom teeth if we have mouths that seem to be too small for them today? Why do humans have an appendix if we don't need it? Levels of organization within the body are also important. Ecosystems are made up of organisms that are made from organ systems. Organ systems are made from organs that are made of tissues. Tissues are made of cells. These cells are the building blocks of all living things. They are known as the basic units of structure and function.

In "Structure and Function Creatures," students design and build creatures that have many appendages. Each appendage or other part of the body will have a function or a purpose (reason) for being there.

In "Onions Versus Whitefish," the plants take on the animals in a battle to the end. In a debate format, students take sides and argue why plant cells are better than animal cells and vice versa. Misconception alert! Be sure your students realize that both plant cells and animal cells have cell membranes. Many students believe that plant cells only have a cell wall and no membrane.

Structure and Function Creatures

SCIENCE TOPICS

structure, function, habitat, adaptation

EDUCATIONAL GOALS

Students will

- relate structure to function in the organism as a whole.
- relate cell structure and function to the higher levels of organization including tissues, organs, organ systems, organisms, and ecosystems.
- show understanding of adaptations.

MULTIPLE INTELLIGENCES

logical/mathematical, verbal/linguistic, bodily/kinesthetic, musical/rhythmic, visual/spatial, interpersonal, and naturalist

MATERIALS

construction paper, drawing paper, crayons, markers, paint sets

PROCEDURE

Students will work in groups of three to create an imaginary structure/function creature. They will draw their creatures and present them to the other members of the class.

DIFFERENTIATION

The lesson is tiered, since students can choose the complexity of structures and functions to include in their project. The products will naturally be tiered as a result. Although this is assigned as a small-group activity, individuals will each be responsible for designing ten structures and functions that do not overlap with the others. Students will have full choice in the types of structure/function creatures they create.

ASSESSMENT

Students will be graded on the amount of research they include in their project. They will also be graded individually to be sure each member of the group has ten structures and functions. The thirty structures and functions should be listed on a sheet of paper that can be used during the oral presentation to the class. By assigning such a large number of structures and functions per group, it will push students to be creative and go beyond the traditional options. The behaviors of the organism may also be included. Some students may also choose to explore plantlike or bacterialike organisms in addition to animallike organisms.

Structure and Function Creatures

Plants, animals, and all living organisms have structures that are adaptations to their environments that allow them to survive harsh or extreme conditions. This activity will have you working in a group of three to design your own structure/function creatures. The creature may be fictional or loosely based on a plant, an animal, or another existing organism.

Directions: In a group of three, design an imaginary creature that has thirty different structures and functions. You will list these structures and functions on a sheet of paper and present them to the class. Each member of the group will be responsible for designing a minimum of ten structures and listing their functions for the creature. These structures cannot be the same as ones created by the other group members, so communication is important throughout the design process. In addition to the traditional definition of *structure*, you may wish to include other things about your creature, such as how structure influences its behavior. The more research you perform and detail you can provide about the creature's structures, the better your grade will be. Once the creature has been designed, you and your partners will work together to draw a colorful picture of it to use during the oral presentation. The oral presentation is also a part of your grade. If you would like, you may also choose to draw the picture on a computer or present it through a PowerPoint presentation.

In the space below, jot down notes or make some rough sketches of your creature.

Onions Versus Whitefish

SCIENCE TOPICS

plant cells, animal cells, cell wall, cell membrane, chlorophyll, chloroplast, mitochondria, cellular respiration, photosynthesis, ATP, glucose, vacuoles, cytoplasm, nucleus, differentiation

EDUCATIONAL GOALS

Students will

- list differences in structure between plant and animal cells.
- list differences in function between plant and animal cells.
- perform a scientific debate with research to support facts given.
- discuss specialized cells and how they differ from other cells through differentiation.
- support why cells (plant or animal) are considered the basic units of structure and function in an organism.

MULTIPLE INTELLIGENCES

logical/mathematical, verbal/linguistic, visual/spatial, interpersonal, and intrapersonal

MATERIALS

research materials for studying differences between plant and animal cells in structure and function, timer for debate

DIFFERENTIATION

Students will write research notes in their science lab journals. Whole-class instruction will be the beginning point, but the class will be divided in half while large groups are investigating either plant cells or animal cells. Note that each group must be able to research enough of the "other side" in order to hold valid arguments for their own side. This will encourage more research by all students. For homework, students can perform research and compile arguments independently. Students should be allowed to choose which side they represent; however, if you are teaching a gifted class or a class you wish to further challenge, you may want to have students choose sides to debate and then "trick them" by switching the sides the students will represent. This will cause students to gather even more data and research to convince themselves why they are truly on the "best side." It will give them practice in deciding what points of a debate should be argued. By having students speak during the debate, you are also encouraging any ESL learners to begin gaining courage to speak in class. By having note cards and/or science journals from which to read, students will feel more comfortable with a prepared speech than if called to speak in front of the class while unprepared.

ASSESSMENT

Teachers can informally assess how much assistance students need throughout the research-gathering process. Teachers will make a list of all students and make a checklist on which they check "no assistance needed," "some assistance needed," or "much assistance needed" during this phase. Teachers can also assess students individually by grading the science lab journals for data pertinent to the debate. Lastly, teachers can grade the debate itself to ensure that every student has something to say and that it is well managed. Debate rules are found on the Internet for various debate teams across the county. Choose one that fits the time and other restraints for your class.

Onions Versus Whitefish

If your teacher told your class that you were his or her favorite class, how would that make you feel? On the other hand, what if your teacher said he or she has taught for over thirty years and based on test grades gathered, behavior charts, and other data, your class was her favorite? The more data or supportive statements that are given to back up an opinion, the more believable the argument. For this reason, research is extremely important in science and debates.

Directions: You will take part in a class debate to determine which cell is superior—the plant cell or the animal cell. You will work with your teacher to decide which side you will be on for the debate. You and your classmates will be studying onion (plant) cells and whitefish (animal) cells. The onion and the whitefish are the two most common organisms observed under the microscope to represent plant and animal cells. Observing these two types of cells will help you gather information for the debate.

In your science lab journal, record all notes you make during the research phase of the debate. Important science concepts/terms you may wish to research may include plant cells, animal cells, cell wall, cell membrane, chlorophyll, chloroplast, mitochondria, cellular respiration, photosynthesis, ATP, glucose, vacuoles, cytoplasm, nucleus, and differentiation. Remember that to have an intelligent argument or debate, you will need to understand both sides of the issue. In other words, you will need to anticipate the debate points that could be made by the other side. This way you can prepare for anything the other side may decide to throw your way. Good luck!

Life Science
Reproduction and Heredity

Reproduction is essential for the survival of a species, but not an individual organism. Some organisms reproduce sexually, whereas others reproduce asexually through cloning and other mechanisms. Sexually produced offspring are never identical to their parents due to genetic variation that occurs during the crossover phase of meiosis.

Hereditary traits are passed from one generation to another through chromosomes. Genes on the chromosomes can determine one or more characteristics in the offspring. Some traits are determined by the genes; however, some are determined by the environment of the individual. This point is often argued by the "nature versus nurture" debate.

"Extracting DNA from Cheek Cells" allows students to obtain their DNA with simple household items. In addition, DNA from plants such as bananas or onions can also be obtained by the same methods. This is a good lab to do after you have already had students examine their cheek cells with dye under the microscope after making scrapings with a toothpick.

"Protein Manufacturing" gets students on their feet when they become DNA, mRNA, tRNA, and amino acids in order to build a protein.

Extracting DNA from Cheek Cells

SCIENCE TOPICS

DNA, heredity, science process skills

EDUCATIONAL GOALS

Students will

- extract DNA out of a living cell.
- list the steps to isolate DNA from cells.
- observe DNA strands under the microscope.
- compare and contrast human DNA to plant DNA strands (optional).

MULTIPLE INTELLIGENCES

logical/mathematical, bodily/kinesthetic, visual/spatial, interpersonal, and naturalist

MATERIALS

3-oz. paper cups, salt, test tubes, liquid dish detergent, ice-cold ethanol (denatured alcohol) or ice-cold isopropyl alcohol (rubbing alcohol), goggles, apron, thin glass stirring rod (optional), banana (optional), and onion (optional)

SAFETY NOTE

When isolating human DNA, sterile procedures should be taken to reduce the spread of disease through the saliva. A teacher may also choose to only do the plant DNA and skip the human DNA. Goggles should also be worn to reduce the chances of alcohol getting into the eyes.

DIFFERENTIATION

More advanced students may be able to obtain their DNA and then move on to isolate the DNA from both the banana and the onion. Advanced students could also isolate DNA at home for homework, trying various other fruits and vegetables, and report their findings to the class. The lab itself begins with whole-class instruction and then utilizes small-group instruction while the lab is being conducted. Lab notes will be made on an individual basis in the student lab journals.

ASSESSMENT

Students will be graded on their student lab journals. Comparisons should be made (if performing the optional portions of the lab) between the various DNA extractions.

Extracting DNA from Cheek Cells

Directions: Follow the steps below using the materials your teacher provides. Make sure you follow any safety procedures your teacher specifies. Then answer the questions that follow.

1. Wear goggles and an apron. Swish about 30 mL of 0.9% salt water in your mouth for 30 seconds. Swish vigorously! This will be hard work, so be ready. You are attempting to loosen as many cheek cells into the salt water as you possibly can.

2. Spit the water into your paper cup. Pour this into a test tube containing 5 mL of 25% liquid dish detergent water.

3. Rock the test tube gently on its side for two to three minutes. The detergent will break open the cell membrane and release the DNA into the soap solution. Don't be too vigorous while mixing! DNA is a very long molecule. Physical abuse can break it into smaller fragments.

4. Tilt the tube and pour 5 mL of the chilled 95% ethanol down the side so it forms a layer on the top of your soapy solution.

5. Allow the tube to stand for one minute.

6. Place a thin glass stirring rod into your test tube.

7. Stir or twirl the rod in one direction to wind the DNA strands onto the rod. Be careful to minimize mixing of the ethanol and soapy layers. If too much breaking of the DNA has occurred, the DNA fragments may be too short to wind up, and they may form clumps instead. You can try to scrape these out with the rod. *Hint:* The DNA will be easier to spool onto the rod if the glass is freshly cleaned and dry.

8. After you have wrapped as much DNA onto the rod as you can, you may wash it down the sink with water. Wash all equipment with soapy water to disinfect it. Be sure to leave your lab station as clean as when you got there.

(continued)

Extracting DNA from Cheek Cells (continued)

Answer the following questions below or in your science lab journal.

1. Did you isolate your DNA? Why or why not?

2. Why did you swish your mouth with salt solution?

3. Where were the strands of DNA first observed?

4. Why was it important to spin the stirring rod in only one direction?

5. What does the detergent do in the solution?

6. Why do you think the alcohol needs to be cold? Do you think the lab would work with other types of alcohol? Why or why not?

7. (Optional) If you also isolated DNA from plants, how did it compare to your DNA?

Reproduction and Heredity
Differentiated Instruction for Science

Protein Manufacturing

SCIENCE TOPICS

DNA, nucleus, ribosome, mRNA, tRNA, protein, transcription, translation, replication, amino acids, codon, protein, adenine, thymine, guanine, cytosine, uracil

EDUCATIONAL GOALS

Students will

- list the steps of replication, transcription, and translation including all the DNA and RNA bases.
- locate the parts within the cell where replication, transcription, and translation occur.

MULTIPLE INTELLIGENCES

logical/mathematical, verbal/linguistic, bodily/kinesthetic, musical/rhythmic, visual/spatial, interpersonal, intrapersonal, and naturalist

MATERIALS

yarn or string, paper, DNA music (optional; available on the Internet at toddbarton.com).

PROCEDURE

1. Make signs on paper that read adenine (A), guanine (G), cytosine (C), thymine (T), and uracil (U). Make larger signs that read replication, translation, transcription, nucleus, ribosome, protein, codon, mRNA, DNA, tRNA, protein, and amino acid. Punch a hole in the top of each sign and put the yarn/string through the signs to make a "necklace" out of each. If you have more than fifteen students, continue making sets of the smaller signs with adenine, guanine, cytosine, thymine, and uracil on them. In a smaller class, it would also be possible for some of the students to move from one area of the "cell" to the next with a reminder that it does not actually happen this way in an actual cell. Place the signs face down on a lab counter or table for students to pick up as they enter the room. The activity can be conducted outdoors if desired.

2. Students with the replication, transcription, nucleus, translation, and ribosome signs will line up first (with large spaces in between them) in the order listed. These will be the areas where those processes occur. Remind students that replication and transcription occur in the nucleus of the cell, but translation occurs at the ribosome of the cell. The student with the DNA sign will stay in the nucleus of the cell, while the mRNA student will move from the nucleus to the ribosome between the steps of transcription and translation.

3. Students who have the DNA and RNA bases will line up randomly, beginning with those occurring in DNA only. The process of replication should be shown as students hold hands to show bases joining in the pairs A-T and G-C or T-A and C-G. Remind students that chemically it is impossible for A-G, A-C, and so forth to form. Replication occurs so that if DNA becomes damaged, there is another copy of it. In other words, DNA makes a copy of itself before the protein manufacturing process begins.

4. Next, students will show the process of transcription. Transcription is the process in which mRNA (messenger RNA) comes in to make a copy of the DNA. Once the students representing the bases of mRNA come in, remind them that uracil is a part of mRNA instead of thymine. Now, the pairings will be A-U or U-A and G-C or C-G.

5. Now, translation occurs. Translation (as the name implies) occurs when tRNA transfers the message from the mRNA molecule and it is translated into the sequence of amino acids forming the protein being made. The bases are read in sets of three bases known as codons. This "code" is read within the ribosome of the cell. The tRNA gets the amino acid coded by the codon or set of bases. Peptide bonds form between the series of amino acids, and a long chain of protein develops.

DIFFERENTIATION

Students will write notes during the process in their science lab journals. The activity is tiered, with students playing various roles throughout the experiment. As a result, students will produce tiered products to demonstrate understanding of the concepts. This problem-based activity is a whole-class experiment.

ASSESSMENT

Students will list the stages of replication, transcription, and translation on the activity sheet or in their science lab journals. Descriptions can be graded for accuracy. Science terms listed should be included in their description. Questions about this activity can be asked to test students' knowledge.

Protein Manufacturing

In this lab, you will become a part of the process of making a protein in a cell. Use the materials provided by your teacher, and follow your teacher's instructions. Make notes and record in your science lab journal any observations you make.

Once you have completed the lab, answer the questions on the lines below or on a separate sheet of paper in your science lab journal.

1. Describe the process of replication. Where does it occur?

   _____
   _____
   _____

2. Describe the process of transcription. Where does it occur?

   _____
   _____
   _____

3. Describe the process of translation. Where does it occur?

   _____
   _____
   _____

4. What types of bonds hold the amino acids of a protein together?

   _____

5. What part did you play? What would happen to a protein (end product) if your part were missing?

   _____
   _____
   _____

6. Why does DNA need to make a copy of itself before the entire process begins?

   _____
   _____
   _____

Reproduction and Heredity
Differentiated Instruction for Science

Life Science
Regulation and Behavior

Behavior is regulated through the nervous system. It is also related to an organism's adaptations to its surroundings. Examples of an organism's behavior can include how it gets its food, responds to emergencies, moves, and reproduces. Part of the behavior is determined by genetics, while part of it may be determined by environmental factors. The acts of protecting their young and hunting are also instinctive behaviors in animals. Remind students that animal behavior also includes things such as migrating and hibernating.

In "How Much Is That Ficus in the Window?" the focus shifts from a childhood song "How Much Is That Doggie in the Window?" to plants and plant behavior. When most students think about behavior of organisms, they think about human or mammal behavior, forgetting that plants have behaviors, too. This lab may be extended to include behavior of other organisms such as earthworms, frogs, and so forth. Just be sure if you do so that no organisms are harmed and that they are placed back into their natural environments.

In "Feeling Blue Today?" human psychology is explored in terms of an organism's response to its environment. The nervous system controls this response. This is also a good activity to draw attention to how difficult habits are to break.

How Much Is That Ficus in the Window?

SCIENCE TOPICS

tropisms, plant behavior, chemotropism, phototropism, geotropism, hydrotropism, thigmotropism, auxin

EDUCATIONAL GOALS

Students will

- describe the various tropisms in plants through dance and song.
- relate the tropisms of plants to plant behavior and compare them to animal behavior.
- analyze various behaviors in nature as a result of a walk through the woods.

MULTIPLE INTELLIGENCES

logical/mathematical, verbal/linguistic, bodily/kinesthetic, musical/rhythmic, visual/spatial, interpersonal, intrapersonal, and naturalist

MATERIALS

CD player (optional), a CD with a copy of the song "How Much Is That Doggie in the Window?" (optional)

PROCEDURE

Take students outdoors on a nature hike. They need to write notes in their science lab journals about any plant or animal behaviors they see. When students return to the classroom, have some share a few of the things they noted. Did any note plant behaviors? Although plants move slowly, they have behaviors, too. Maybe you live in a part of the country that has kudzu, which grows an average of a foot a day. Possibly, you have grapes, ivy, or other vines that curl around things. Maybe you have Spanish moss hanging from the trees.

After students have had an opportunity to share, write the various tropisms on the board or use a PowerPoint slide. Tell students that the word *tropos* is Greek for "turn." An extension activity of this lab would be to take time to grow plants under various conditions or have students do this at home for homework and describe what they see.

The words and definitions used in this activity are as follows:

- chemotropism: plant response to chemicals
- phototropism: plant response to light
- hydrotropism: plant response to water or moisture
- thigmotropism: plant response to touch
- geotropism: plant response to gravity
- tropos: Greek for "turn"

Each of the tropisms is in response to a certain thing. The tropism may be a negative or a positive response. For example, when a plant's roots grow into the ground in response to gravity, that is a positive response. When a plant's stems, trunk, and branches grow away from the downward pull of gravity (as they should), then that is a negative response.

Teach the class the melody of "How Much Is That Doggie in the Window?" in case they have forgotten or have never heard the tune before. Then teach them the tropism song titled "How Much Is That Ficus in the Window?" A ficus tree is an ornamental fig tree commonly grown indoors. You may wish to show them a photo of one or find an Internet site with pictures of one. You may want to make motions or movements to each of the stanzas; for instance, bend when you sing about the bend in the trunk and so forth.

DIFFERENTIATION

This activity utilizes whole-class instruction as students walk in the woods together to note various behaviors. However, each student's experience will be different due to the unique observations each student will make. Allowing those students who wish to express themselves through art or poetry would be a method for differentiating this activity.

ASSESSMENT

Both formal and informal assessments through observation would be appropriate for this whole-class activity. The students could grow plants under various conditions and report on their experiments for a grade. Be sure to remind students to have more than one plant when doing a science experiment in case one plant dies.

How Much Is That Ficus in the Window?

Directions: Follow your teacher's directions. As you take a nature hike, you will be looking for animal and plant behavior. Record all your observations in your science lab journal.

The song below can help you remember different tropisms. Your teacher will teach you the song so you can sing it with your classmates.

How Much Is That Ficus in the Window?

How much is that ficus in the window?

The one with the bend in its trunk?

How much is that ficus in the window?

Phototropism from the light caused its ail.

How much is that ivy in the window?

The one with the tendrils so frail?

How much is that ivy in the window?

Thigmotropism from touch wrapped on the rail.

How much is that seedling in the window?

The one with the planter upside down?

How much is that seedling in the window?

Geotropism from gravity made it well.

How much is that palm tree in the window?

The one drooping on its side?

How much is that palm tree in the window?

Hydrotropism (lack of water) made it fall.

Feeling Blue Today?

SCIENCE TOPICS

the nervous system, behavior, psychology, response, stimulus

EDUCATIONAL GOALS

Students will

- relate habits and behaviors to nervous system control and response to the environment.
- define stimulus and response and give examples of each in nature.
- make comparison charts of data and be able to make correct interpretations.

MULTIPLE INTELLIGENCES

logical/mathematical, verbal/linguistic, bodily/kinesthetic, visual/spatial, interpersonal, and intrapersonal

MATERIALS

stimulus/response cards, markers or crayons, timers

PROCEDURE

Teachers first describe a "stimulus" and a "response" to the students. In this scenario, the stimulus consists of cards that are shown to each set of students in class. The response, or "answers" will be given by the students. Teachers can make copies of the cards and allow students to work in pairs to color the cards. Students will make two sets of cards, so they will need one set of cards per two student lab partners. The first set will be colored (or printed on a color printer) to be exactly what the color says. For example, the word blue will be colored blue. The second set of cards made, however, will be different. Although the card itself reads "blue," the students will actually be coloring the inside of the word with a red (or another) color. For the second set of cards, to allow flexibility and differentiation, teachers should allow students to use any color they wish for the inside of the word, as long as it doesn't match what the word itself says.

DIFFERENTIATION

This will be an excellent activity for an ESL student. Colors are usually some of the first words learned in a new language. Not only are you reinforcing the psychology behind the activity, but you are reinforcing the words in English that may be important to certain learners. Allow students to choose their own colors to color in the words on the cards. Do certain sets of cards take longer for the students to use? They can try out several sets and graph the results to compare. Part of this activity will be whole-class discussion during the introduction. Then it will utilize small-group instruction as students work with partners or in small groups. Lastly, students will make reflections in their science journals. For enrichment, data from several classes can be added together for interpretation.

ASSESSMENT

Informal observations can be made during the activity itself regarding what each of the lab partners is doing. Also, science lab journals with reflections from the activity can be graded. Lastly, if students graph results from several sets of cards, their interpretations of the data can be graded.

Feeling Blue Today?

CARD SETS

BLUE

RED

YELLOW

PURPLE

GREEN

(continued)

Feeling Blue Today?

ORANGE
PINK
BLACK
WHITE
BROWN

Feeling Blue Today?

Directions: Follow the steps below using the materials provided by your teacher.

1. You will be given two sets of cards. For the first set of cards, color each card to match its name. For instance, color the word *blue* with a blue color and each of the other colors the "correct" color for the name.

2. For the second set of cards, however, color each of the cards a color different from the word actually stated on the card. For example, the first word card has the word *blue,* so you will color it another color such as red.

3. Once all the card sets have been colored, time yourself saying the correct colors of the words in card set 1. Record the time in your science lab journal. Repeat twice more and calculate an average time.

4. Time yourself saying the correct *colors* of the words in card set 2. Record the time in your science lab journal. Repeat twice more and calculate an average time.

5. Then time yourself reading the *words* of card set 2 (regardless of the color on the inside of each word). Record the time in your science lab journal. Repeat twice more and calculate an average time.

6. Make a graph of the data for each of the trials.

7. Reflect in your lab journal about each of the three activities above based on the data you have collected.

8. If your teacher directs you to do so, trade card set 2 with another group of students in class and compare to your own set. Reflect on the similarities and differences.

Life Science
Populations and Ecosystems

A population is defined as groups of organisms of the same species living in the same place at the same time. The ecosystem is made up of the populations of various species (community) in a location along with the nonliving parts of that location. The nonliving parts would include soil, rocks, air, water, temperature, and sunlight. The living parts are categorized into one of three areas: producers, consumers, and decomposers.

"Measuring Up to a Blue Whale" is a problem-based lab in which students have become curators of a new, large aquarium opening in their city. They will need to build a tank large enough to hold a blue whale. But how large is that? Students will explore the meaning of *large* in this activity. An introduction to food pyramids and biomass levels is included. The students will easily see why the largest animals on Earth are herbivores.

"Don't Get Caught in the Food Web!" physically shows students how a food web works and what happens if one of the members is taken out. It clarifies for students the roles of producers, consumers, and decomposers. Human impact on ecosystems is explored in this activity.

Measuring Up to a Blue Whale

SCIENCE TOPICS

measurement, estimation, ratios, proportions, mammals, blue whales

EDUCATIONAL GOALS

Students will

- estimate measurements such as height and weight.

- apply mathematical formulas of proportions to make more accurate estimations.

- outline a blue whale's length and describe its weight, food sources, and behavior.

- reflect on what a blue whale sounds like.

MULTIPLE INTELLIGENCES

logical/mathematical, verbal/linguistic, bodily/kinesthetic, musical/rhythmic, visual/spatial, interpersonal, intrapersonal, and naturalist

MATERIALS

metric tape measure, rulers, masking tape, blue whale "songs" to play during activity (also available for download on the Internet at whalesounds.com).
Note: Most whale sounds are taped from the humpback whale. If you do a search, however, you can find small clips of the clicking sounds made by blue whales. Blue whales make the loudest sounds of all mammals at 188 decibels, which is louder than a jet engine!

PROCEDURE

Play whale songs/music while students work. Students will make estimations of adult blue whale length and weight and then perform research to find out the actual numbers. You will then give them directions about measuring and taping an area to represent an adult blue whale. If you prefer, you may be able to obtain crime scene or warning yellow tape to span the distance so it will not be too sticky. It could be taped down every few meters to keep it from moving. As students are working, remind them that the largest animals on Earth usually eat plants, small animals such as krill, and plankton because they are able to obtain more energy from these types of food sources than if they were strictly carnivores or meat-eaters.

DIFFERENTIATION

In this problem-based lab, students will work in groups to determine how long a blue whale measures. They will do mathematical formulas with proportions to find out how whales compare with themselves. Students will also listen to the sounds of whales communicating. The beginning of the activity is a whole-group discussion, but then the activity becomes a group activity. To differentiate further, other large mammals could be explored in this same manner, and students could choose which ones to work on in smaller groups. Some possibilities are giraffes, hippopotamuses, elephants, bears, bison, and so forth.

EXTENSION ACTIVITIES

More advanced students may want to build a life-size model of a whale out of black trash bags or black painter tarps/plastic that have been taped together. The whale shape should be three-dimensional so students can crawl inside the whale. This would also be a great way to pull in the community as well. To create an entrance, you may need to run a fan at either the tail side of the whale or the mouth side so students can crawl inside.

Have students research on their own various aquariums located throughout the world. What are the largest species contained? How much water do the various aquariums hold? How large are the aquariums when you measure length, width, and height? How much longer/shorter is that than the size of an adult blue whale?

ASSESSMENT

Students will record data as they go through the process of finding a tank large enough for a blue whale. They will also reflect on how easy or difficult the task was. Students will write how they felt when they realized how large a blue whale actually is. Students should have done a ratio to determine how long a blue whale is, measured in their height. In other words, assuming a student is 5 feet tall and a whale is an average 100 feet long, a whale would be 20 students long!

Measuring Up to a Blue Whale

Directions: A new aquarium is being built in your community. Congratulations! You have now been named the curator for the new aquarium. You have just one problem. In the history of aquariums throughout the world, no one has been able to house very large fish or mammals (such as the whale). It is up to you to find out exactly how large a blue whale is and to decide whether it is possible to house such an animal in your aquarium.

The first step is to make a hypothesis, or educated guess, about the size of a blue whale. Record your hypothesis below. How many feet (meters) do you think an adult blue whale measures? How many pounds do you think an adult blue whale weighs?

_____

_____

Conduct research of your own to determine the actual length and width of a blue whale. Record your findings below.

_____

_____

Once your teacher gives you directions to do so, you will be taping off part of your school to show the size of a blue whale. It will take your entire class working together to accomplish this.

After the activity is completed, answer the following questions in your science lab journal.

1. How easy or difficult did you feel the task was before you started?

2. How did you feel when you realized how large a blue whale actually is?

3. How tall are you in feet? in meters? How does your height compare to the length of a whale?

4. Find out how long a whale is based on your height. For instance, how many of you would equal one whale? You will need to do proportions to figure out the answer to this problem.

Don't Get Caught in the Food Web!

SCIENCE TOPICS

food chain, food web, producer, consumer, decomposer

EDUCATIONAL GOALS

Students will

- compare and contrast food webs and food chains.
- compare and contrast producers, consumers, and decomposers.
- describe what occurs when a complex food web is broken by humans affecting their ecosystems.

MULTIPLE INTELLIGENCES

logical/mathematical, verbal/linguistic, bodily/kinesthetic, visual/spatial, interpersonal, intrapersonal, and naturalist

MATERIALS

eight or more skeins of different colors of yarn; resource materials on food chains, producers, consumers, and decomposers; pieces of paper; markers or crayons

PROCEDURE

In groups, students will form food chains. Each student will be one link in the food chain. Each student will write his or her part on a piece of paper that will hang as a sign around the student's neck. Then, groups will come together to form food webs according to the directions in the next step. Here are two examples of possible food chains:

- dandelion, deer, lion, hyena, blowfly, and mushroom
- grass, caterpillar, cardinal, hawk, vulture, and millipede

Note that the first organism in each example is a producer, or an organism that makes its own food. Bacteria and one-celled water organisms can fit there, too. The next level consists of various consumers including herbivores (plant-eaters), carnivores (meat-eaters), or omnivores (organisms that eat both plants and animals). You may want to explain the concept of primary consumers, secondary consumers, and so on to students. The last level is composed of decomposers. Decomposers feed on dead plants and animals and eat wastes. Examples of other decomposers are centipedes, pill bugs (isopods), earthworms, and mushrooms.

Students representing producers will each take a different colored skein of yarn. The entire class will form a large circle, with everyone mixed together in no particular order. This is a great opportunity to go outdoors to allow enough space. The producers will find where the next level of their particular food chain (group partner) is and throw the yarn to them. The yarn will continue to be thrown until all members of that particular chain have yarn attaching them together. The teacher will now talk about food chains versus food webs while the students note what is being said based on the strings attaching them. Students will note what happens when humans have affected the food web by something such as spraying pesticides if the teacher decides to cut certain strings or instructs students to drop their strings to the ground to break the connections.

The second part of this activity continues straight from the first. Students stay in the same place. Now, students have a chance to be creative with their food webs. They can choose who to throw their yarn to as long as it makes sense on a food chain. New and limitless combinations are now possible. The teacher may want to challenge the students to see how many students they can possibly involve in the food web and how complex they can make it. Once again, however, the strings need to be cut or dropped to show what an impact something can have on a stable ecosystem. A disease could even be the cause of an unstable ecosystem.

DIFFERENTIATION

This activity will begin as a jigsaw activity in which groups of three to four students will work together to create a food chain. Once the food chains are created, each group will join the other small groups for a whole-group activity. Students will have choices about which parts of the food chain to be. They will reflect on the activity in their science lab journals.

ASSESSMENT

Teachers will grade students' answers to the questions and reflections on the activity. Informal assessment made during the activity itself can also be included.

Don't Get Caught in the Food Web!

Directions: Work with two or three other students in a "food chain" group according to your teacher's directions. Research the terms below and write their definitions on the lines provided. For each term, include at least one example. (For instance, grass is an example of a producer.)

1. producer _____

2. consumer _____

3. decomposer _____

Together, these three things make a food chain. Now, it is your turn to form a food chain with the other members of your group. For instance, one person can be a producer, one person can be a consumer, and so forth. Just make sure your group has all the parts necessary to form a food chain.

Now, write your role in the food chain on a piece of paper. Your teacher will give you a piece of yarn. Punch a hole in the top of your paper and string your yarn through it. You will wear your sign around your neck for the next part of the activity.

Follow your teacher's instructions to create a food web with the rest of your classmates. Each student should have a piece of paper that identifies him or her as a producer, a consumer, or a decomposer.

Once you and your classmates have finished the activity, answer the following questions on the lines below or in your science lab journal.

4. What is the difference between a food web and a food chain?
   _____
   _____

5. What are two examples of things that humans do that can affect a food web in a negative way?
   _____
   _____

6. Did the group activity help you understand how food chains work? Why or why not?
   _____
   _____

Life Science
Diversity and Adaptations of Organisms

This section of the book is an excellent follow-up to the section addressing structure and function within living systems. Many species of organisms are alive on Earth today and have surprising similarities. With the advent of the DNA Human Genome Project in addition to new studies in genetics and DNA with plants and other organisms, we are discovering how similar we all are. Through this same technology, we are discovering that we have placed some organisms incorrectly when attempting to classify them into taxonomic groups.

Evolution (a gradual change over a very long time) is science's best theory as to why organisms have similarities to and differences from one another. As students developed structure/function creatures in Activity 10, they were able to see some imaginary adaptations (changes in behavior, physiology, or anatomy) that may have helped their organism survive and possibly even reproduce. In our real world, many adaptations have occurred, but most of them happened from random, accidental genetic mutation. Once this mutation occurred, then it was possible for that organism to survive and possibly thrive and pass on its genes. Often, teachers confuse this theory and state that an animal changes to adapt to its environment and then passes on its genes. If an organism does not survive after a mutation or a change in environment, then it may die. If large numbers of one species die, they become endangered. If all of one species die, such as happened with the dinosaurs or dodo birds, then they are called extinct. Extinction is a common occurrence.

In "Giraffes Can't Jump!" students will apply math concepts such as graphing and measuring, which overlap many science-process skills such as measuring and averaging. Graphing will assist the students in drawing conclusions from the data gathered during the experiment.

"Extincting Extinctions" is a problem-based, research-based activity in which students problem-solve to limit the number of extinctions that are sure to occur because of human interactions with ecosystems. Students are given choices about what to research and about how to present their findings to the class when finished.

Giraffes Can't Jump!

Adapted from Krista Parker, Woodward Academy, GA

SCIENCE TOPICS

natural selection, survival of the fittest, natural disasters, measurement, averaging

EDUCATIONAL GOALS

Students will

- graph and plot heights of student "giraffes" four times in a class period.
- make predictions about giraffe survivability and characteristics after each natural disaster.
- be able to calculate an average.
- measure using the metric system.
- understand that most mutations occur randomly and by chance and then nature "selects" them so that they are expressed in the next generation.

MULTIPLE INTELLIGENCES

logical/mathematical, verbal/linguistic, bodily/kinesthetic, visual/spatial, interpersonal, intrapersonal, and naturalist

MATERIALS

60 construction paper leaves that measure approximately 20 cm. in length and 10 cm. in width, metersticks, graph paper, a place to record data for the class (such as on the board or on the computer)

PROCEDURES

Clear space in your classroom or go outside or to an open space for this activity. Choose four students of average height to be "trees." *Safety note:* The "trees" will need to stand on a short chair to be taller than most of the students or "giraffes." Giraffes may need to walk on their knees if safety issues seem to arise. The four trees each hold five leaves fanned out in each of their hands. Their hands are held at a 90-degree angle from their body. The rest of the leaves are scattered on the ground. All the remaining students become giraffes. The giraffes all need to record their height in centimeters in their science journals and on the board/computer for a giraffe height average to be calculated and later graphed. Remind students that giraffes cannot jump, push, or talk! Giraffes are allowed to gather leaves for 20 seconds or until all the leaves are gone. Any giraffe that does not have five or more leaves is dead due to starvation. The heights of surviving giraffes are listed in student science journals and on the board/computer for a new giraffe average height.

Now, there has been a drought in the forest. Due to this, there are no longer any leaves on the ground. Leaves are only in the trees, and the trees now have their "branches" held up at a 45-degree angle, still with only five leaves in each hand. The extra leaves have been put away. Once again, remaining giraffes have 20 seconds to gather leaves, and they will die without at least five. Heights of the "living" giraffes are recorded on the board/computer again in order to calculate height averages.

Lastly, there is a forest fire in the woods. Now, there are no leaves under the trees, and the branches are straight up against the side of the trunk (ears) of the trees. Each hand holds five leaves. Surviving giraffes once again have 20 seconds to attempt to gather five leaves from the trees with the reminder that "giraffes can't jump." One last graph is made and then averages are taken.

DIFFERENTIATION

The numbers of leaves required for the "giraffes" to survive may be changed so that comparisons can be made. There can be more or fewer leaves at the beginning of this problem-based activity in order to see how that affects the outcome. If you are working with more advanced students, more variables may be added to the experiment. For instance, blue water drops made of construction paper may be added to simulate water, which is also necessary to sustain life. Once again, without enough water (even with enough food) life will not continue. The more variables there are, the more difficult the graphing will become. Another variation is for students to use graphing spreadsheets on the computer. Students will record data in their science journals and reflect on the activity as well.

ASSESSMENT

1. Assess class participation for giraffes and trees.
2. Grade journals with average giraffe height calculations.
3. Review reflections from the lesson.

Diversity and Adaptations of Organisms
Differentiated Instruction for Science

Giraffes Can't Jump!

Adapted from Krista Parker, Woodward Academy, GA

Directions: Follow your teacher's directions to see if you will be a giraffe or a tree. Record data as it is written on the board or computer into your science lab journal so you will be able to make interpretations later.

Be sure to graph all data into a spreadsheet on the computer or onto graph paper, which will make it easier to draw conclusions at the end of the lab.

Once you have finished the lab, answer the following on the lines below or on a separate sheet of paper in your science lab journal.

1. Based on the number of giraffes that survived after each of the natural disasters in the lab, predict what would happen if a natural disaster occurred to a stable giraffe population.

   _____
   _____
   _____
   _____

2. Name some natural disasters that have affected animal populations.

   _____
   _____
   _____
   _____

3. Discuss some animal adaptations that were evident after one of the natural disasters you named above.

   _____
   _____
   _____
   _____
   _____

Extincting Extinctions

SCIENCE TOPICS

endangered species, extinction, adaptation, evolution, ecosystem

EDUCATIONAL GOALS

Students will

- research conditions (environmental, genetic, and other) that cause organisms to become endangered and then extinct.

- choose one extinct organism and research the possible reasons why it became extinct.

- rationalize what humans could have done to save the life of this particular organism (if humans existed while this organism was living). If humans were not yet in existence, hypothetically what could have been done to prevent its extinction?

- list programs and processes humans employ today to prevent extinction of species on Earth.

- present research about the extinct organism to the class in the form of an oral report with a typed paper, a PowerPoint presentation, a poster or other visual with discussion, detailed artwork of the organism with supporting presentation, student-produced videotape, or other teacher-approved product.

MULTIPLE INTELLIGENCES

logical/mathematical, verbal/linguistic, bodily/kinesthetic, musical/rhythmic, visual/spatial, interpersonal, intrapersonal, and naturalist

MATERIALS

resource and reference materials, various presentation materials, arts and craft supplies, scissors, tape, markers and crayons, CD player, VCR

DIFFERENTIATION

Students will be able to choose the extinct organism to research in this problem-based, tiered lesson. There will be a variety of materials utilized by each student depending on his or her choice of the finished product. The class will begin with whole-class instruction, but will move into independent study once projects are

chosen. Homework will also be assigned according to individual project needs. Research materials will be placed in interest centers for student use.

ASSESSMENT

The teacher and students will work together to define the levels of a good project and develop a rubric before the research for the projects begins. The rubric must be general enough to cover various types of products students will be producing.

Extincting Extinctions

Directions: You have just been promoted to be in charge of the United States Fish and Wildlife Service. In this new position, your first area of concern is preserving current wildlife for all citizens in the nation to enjoy for years to come. You worry about hunting, overfishing, pollution, and loss of animal habitat due to human interaction with the ecosystems. In fact, the more you research, you find many organisms are now on the endangered species list. It is not just mammals on the list, however. The list also includes birds, invertebrates, reptiles, plants, amphibians, and fish. Additionally, there are many animals on the threatened list, too.

Your job is to see where humans may have gone wrong in the past. As you know, answers are often found in history. If you look at organisms from the past that have already become extinct, you may realize how to preserve the organisms currently on Earth. Research conditions that have caused species to become extinct. Then choose one extinct species to present to the class. Of course, there are animals that became extinct before humans even existed. If you choose a species from that far back in time, you can still hypothesize about how you feel humans could have helped prevent that extinction.

You will work with your teacher to design a rubric to determine how you will be judged on the project. You will also have a choice in how to present your research to your classmates. You may choose to do an oral report with a typed paper, give a PowerPoint presentation, create a poster or other visual with discussion, create detailed artwork of the organism with supporting presentation, or produce a videotape or other approved product. Remember to support everything you present to the class with facts from your research.

Earth and Space Science
Structure of the Earth System

The atmosphere that surrounds Earth is made up of nitrogen, oxygen, and trace gases including water vapor. Earth itself is also layered and includes the hydrosphere (liquid water and ice), which makes up the liquid portions of Earth's surface, and the lithosphere (Earth's soil and rocks), which makes up the solid portions of Earth's surface.

Furthermore, the lithosphere is also layered. There are four major layers: the outermost layer, the crust; the mantle; the outer core that is liquid iron and nickel; and the inner core that is solid iron and nickel.

Several web sites offer simulations and lesson plans for earth science including discoverourearth.org. This site includes content on continental drift, plate boundaries, seafloor spreading, earthquake depth, earthquake magnitude, volcanoes, and topography.

"The Water Cycle Song" uses a song based on a children's tune to help students remember the steps of the water cycle. The activity also helps to remind students that a cycle is a circular pattern containing steps in a particular order. In groups, students will write their own songs about the water cycle and create motions or dance steps to go along with them.

"Rocks Bingo" is a variation on the traditional bingo game in which vocabulary words are substituted for numbers and letters. This engaging game may be applied to many different topics and subject areas. Vocabulary words can be changed depending on the topic being introduced or reviewed.

"Cloud Cover Song" features a song written by meteorologist Nick Walker, otherwise known as the "Weather Dude." The song teaches students about cloud types, including how the shapes and colors of clouds can help predict various types of weather. In groups, students will create their own songs about cloud types.

The Water Cycle Song

SCIENCE TOPICS

precipitation, condensation, evaporation, clouds, accumulation, the water cycle

EDUCATIONAL GOALS

Students will

- put the steps of the water cycle in the proper order.
- sing a song about the water cycle with their classmates.
- write and perform their own song about the water cycle.

MULTIPLE INTELLIGENCES

logical/mathematical, verbal/linguistic, bodily/kinesthetic, musical/rhythmic, visual/spatial, interpersonal, and intrapersonal

MATERIALS

CD player (optional)

DIFFERENTIATION

Although the lab begins as a whole-class activity, it becomes a small-group activity. This is a good activity for students with learning disabilities or those who are English language learners because the teacher first shows students what to do and then students mimic the activity. Students will eventually work together to create their own songs and movements to accompany the songs. This activity allows a great deal of student choice and encourages students to be as creative as possible. After songs are written, students will reflect on the experience in their science lab journals.

ASSESSMENT

Songs that students create can be graded for originality and accuracy. Reflections from the activity can also be graded.

Structure of the Earth System
Differentiated Instruction for Science

The Water Cycle Song

Sing "The Water Cycle Song" along with your teacher. Once you have the idea, you and a group of three other students will work together to write your own song about the water cycle complete with motions.

The Water Cycle Song

by Dawn Hudson

(Sung to the tune of "Row, Row, Row, Your Boat")

Rain, rain, rain at last! Precipitation is its name!
Merrily! Merrily! Merrily! Merrily! Evaporation is to blame.

Clouds, clouds, clouds in the sky! Evaporation caused you to form!
Merrily! Merrily! Merrily! Merrily! Condensation is the norm.

Lakes, rivers, streams, and oceans! A place from which to condense!
Merrily! Merrily! Merrily! Merrily! Accumulation's not nonsense.

Water cycle's a circle! Round and round 'tis true!
Pick it up where you want, it'll stick in your mind like glue.

Evaporation! Condensation! Two more steps to go!
Precipitation! Accumulation! Now, let's wait for snow.

Now, it's your turn to write your own song about the water cycle. Be sure you have motions or a dance to accompany your song. Write the song in your science lab journal. You will be sharing your song with your classmates. When you are finished, be sure to reflect on the work you did and on the work of others in your class.

Rocks Bingo

SCIENCE TOPICS

sedimentary, metamorphic, igneous, various rocks and minerals, properties of rocks and minerals, hardness, cleavage, streak color, luster, fluorescence

EDUCATIONAL GOALS

Students will be able to correctly identify

- minerals found within rocks.
- igneous, sedimentary, and metamorphic rocks.
- rocks and minerals when given a sample.
- properties of rocks and minerals, including hardness, color, cleavage, streak, luster, fluorescence, and so forth when given the proper equipment to measure the properties.

MULTIPLE INTELLIGENCES

logical/mathematical, verbal/linguistic, bodily/kinesthetic, musical/rhythmic, visual/spatial, interpersonal, intrapersonal, and naturalistic

MATERIALS

rock and mineral samples, UV light for fluorescence (optional), nail or penny for scratch test, porcelain plates for streak test, items such as M&M candies to cover ROCKS cards, digital camera (optional)

PROCEDURE

On the board, overhead, or PowerPoint projector, list about 40 words for students to use on their ROCKS cards (minus the free space). Encourage students to use words throughout the list and to mix up their cards as well as they can. Due to having more words than spaces, most cards will be different from one another. Do not allow students to write on the cards once the words are set. You may even want them to use pens so they cannot erase or change their cards, then put the pens away. Students will cover the words that you describe. This gives the activity flexibility since the teacher has control over what is learned from the word list.

Structure of the Earth System
Differentiated Instruction for Science

- **General Rules:** Students cover the word the teacher describes or demonstrates. This is different from traditional bingo in which the actual number to be called is covered. To make the game more interesting, the teacher may begin with traditional bingo in which students get the free space and then need a total of five in a row in any direction to win. Another variation may be allowing students to make the letter *I* in any direction or the letter *T* in any direction. A "small picture frame" would be "bingo" when all the squares around the free space are covered forming a small square. A "large picture frame" would be "bingo" when all the squares around the outside of the card are covered. A "kite" would be four squares in one corner and a "tail" created by the free space and then the next two diagonal squares. "Four corners" would be just the four corners of the card. Feel free to create your own version! To make the game more challenging (and to allow for differentiation), have the winning student call out something other than the words covered to win. (Read examples below.) The winner may get bonus points, recognition, candy, or other prizes.

If students have had a difficult time classifying rocks as sedimentary, igneous, or metamorphic, then use this activity to review that concept. Only use rock/mineral names in the word list. If possible, use a digital camera to project rock/mineral images for the students and have them cover the names of the rocks on their cards. When there is a winner declared, the winner must call out "rocks!" and then specify igneous, metamorphic, or sedimentary for each of the ones they had covered to win.

If students have a difficult time with streak testing or other rock properties, the teacher could call out "produces a reddish-brown streak." This would work as long as the teacher has only one choice in the word list that would fit the description.

DIFFERENTIATION

This activity will utilize whole-class instruction with a tiered lesson possible for the entire class. Various levels of difficulty are possible with the bingo cards since the teacher is able to directly control the level of difficulty of the words to be covered, the definitions to be called out, and how difficult it is for the winner to claim the prize.

ASSESSMENT

Students can be individually assessed as they participate in class. The winners can be assessed as they call back their answers to the teacher.

Rocks Bingo

R	O	C	K	S
		Free Space		

Cloud Cover Song

SCIENCE TOPICS

cirrus, cumulus, and nimbostratus clouds

EDUCATIONAL GOALS

Students will

- sing the cloud cover song with their classmates.
- write and perform their own song about cloud types.
- list the various types of clouds and describe each, including the types of precipitation that may result.

MULTIPLE INTELLIGENCES

logical/mathematical, verbal/linguistic, bodily/kinesthetic, musical/rhythmic, visual/spatial, interpersonal, intrapersonal, and naturalist

MATERIALS

You can find the song "Cloud Cover" by Nick Walker on the Weather Dude web site: wxdude.com. Tune clips from this song and others are also available on the web site.

DIFFERENTIATION

Although the lab begins as a whole-class activity, it becomes a small-group activity. This is a good activity for students with learning disabilities or those who are English language learners because the teacher first shows students what to do and then students mimic the activity. Working in groups, students create their own songs about cloud types. Then they perform their songs for the class and reflect on the experience in their science lab journals.

ASSESSMENT

Students' reflections from the activity can be graded. Original songs can be assessed for creativity and accuracy.

Cloud Cover Song

Sing "Cloud Cover" along with your teacher. Once you have the idea, you and a group of three other students will work together to write your own song about cloud types.

Cloud Cover (Cloud Types)

(Words and Music by Nick Walker. Used with permission.)

This big old ball we call earth has a wondrous atmosphere;
There's oxygen that we breathe in and water vapor here.
Those tiny drops of H_2O climb high till they take flight,
And when they cool they start to pool into fluffs of grey and white.

CHORUS:
I call it Cloud Cover
I keep one eye on the sky.
I'm watching Cloud Cover,
And perhaps you wonder why.
Because when clouds begin to change
They can let the sun shine or bring rain
So watch the Cloud Cover,
Keep a-watching all those clouds.

They're thin and high, those cirrus clouds mean weather's looking fair
Made up of ice, they look so nice like strands of angel hair.
And stratus layers may hang low as a drizzle droplet falls,
Or show me puffs of cumulus like soft white cotton balls.

CHORUS:
I call it Cloud Cover
I keep one eye on the sky.
I'm watching Cloud Cover,
And perhaps you wonder why.
Because when clouds begin to change
They can let the sun shine or bring rain
So watch the Cloud Cover,
Keep a-watching all those clouds.

The nimbostratus might bring rain as a warm front leaves its trail,
And watch with dread as thunderheads bring you lightning storms and hail. (Ouch!)

(continued)

Cloud Cover Song *(continued)*

Cloud Cover
I keep one eye on the sky.
I'm watching Cloud Cover,
Now you know the reason why.
Because when clouds begin to change
They let the sun shine or bring rain
So watch the Cloud Cover,
Keep a-watching all those clouds.

Copyright 1993, 2000 Nick Walker/Faithwalker Music (BMI)

Available from Weather Dude web site: wxdude.com.

Now, it's your turn to write your own song about cloud types. Be sure to include information about the precipitation that results from each cloud type. Work with your partners, and write the song in your science lab journal. You will be sharing your song with your classmates. When you are finished, reflect on the work you did and on the work of others in your class.

Selected Answers

Page 24—Electric Slide

1. $R = 6$ ohms (Ω)
2. $R = 12\ \Omega$
3. $R = 2\ \Omega$
4. $V = 9$

5–7. Answers will vary.

Page 31—Convection, Conduction, and Radiation Station

1. convection
2. convection
3. conduction
4. Answers will vary. Sample answer: To remember the difference between convection and conduction, it can be noted that the word *conduction* has the smaller word *duct* within it. A "duct" is a solid; conduction is heat moving through a solid.

Page 43—Extracting DNA from Cheek Cells

1. Some students may have only isolated plant DNA.
2. to loosen the cells from the cheek
3. floating in solution after the alcohol was added
4. to spool it in one direction since it is a long strand
5. lyses (loosens) or breaks the cell membranes to release the DNA from the cell
6. to keep the DNA from being as disturbed/dissolved; yes, it will work with other types since it is using the OH group from the alcohol to pull the DNA out of solution.
7. Answers will vary.

Page 46—Protein Manufacturing

1. Replication is the process of DNA making a copy of itself. It happens as the double helix "unzips" from the enzymatic action of helicase. At that time, adenine pairs with free-floating thymine (and vice versa), and guanine pairs with free-floating cytosine (and vice versa). This occurs in the nucleus of the cell. The DNA remains in the nucleus to minimize the chances of being destroyed.

2. Transcription is the process of mRNA coming in to make a copy of the DNA in order for protein manufacturing to occur. Unlike DNA, mRNA is single-stranded. It also has uracil instead of thymine. Like the process of replication, transcription occurs in the nucleus of the cell.

3. Translation is the process of mRNA passing the message to tRNA with the use of a ribosome. The genetic code is read by the tRNA molecule. This code is read in sets of three bases called *codons*. The tRNA brings with it amino acids to link together with peptide bonds to eventually form a protein molecule. This process occurs in the cytoplasm of the cell outside of the nucleus.

4. peptide bonds

5. Answers will vary. If a part were missing, it may result in a genetic abnormality due to the deletion of a base and so forth.

6. Replication is important so that nothing happens to the original DNA molecule. DNA is stored in the nucleus of the cell "unwound." As replication occurs, the DNA thickens by the process of making copies. The DNA is more easily damaged while in this state. Ask your students which would be easier to break with one hand—a wet, cooked spaghetti noodle, or one that is dry and uncooked? Most will say the more rigid uncooked noodle. If DNA were always tightly coiled, it would be easier for breakage to occur. By maintaining a more loosely organized state, it is less easily damaged.

Page 59—Measuring Up to a Blue Whale

The average adult blue whale measures 100 feet (30.5 meters) long and weighs an average 150 tons.

Page 62—Don't Get Caught in the Food Web!

Examples will vary.

1. organism that uses the sun's energy to convert carbon dioxide into glucose

2. organism that gets energy by eating producers

3. organism that breaks down dead organic matter

4. A food chain is a series of organisms in which each organism uses the next as a food source; a food web includes all the possible feeding relationships in an ecosystem.

5. Answers will vary. Examples include spraying pesticides, clearing land for new construction, and so forth.

6. Answers will vary.

Bibliography

Differentiated Instruction

Heacox, Diane, Ed.D. *Differentiating Instruction in the Regular Classroom: How to Reach and Teach All Learners, Grades 3–12.* Minneapolis, MN: Free Spirit Publishing, Inc., 2002.

Tomlinson, Carol Ann. *The Differentiated Classroom: Responding to the Needs of All Learners.* Alexandria, VA: Association for Supervision and Curriculum Development, 1999.

Tomlinson, Carol Ann, and Caroline Cunningham Eidson. *Differentiation in Practice: A Resource Guide for Differentiating Curriculum, Grades K–5.* Alexandria, VA: Association for Supervision and Curriculum Development, 2003.

Tomlinson, Carol Ann. *Fulfilling the Promise of the Differentiated Classroom: Strategies and Tools for Responsive Teaching.* Alexandria, VA: Association for Supervision and Curriculum Development, 2003.

Tomlinson, Carol Ann. *How to Differentiate Instruction in Mixed-Ability Classrooms.* 2nd Ed. Alexandria, VA: Association for Supervision and Curriculum Development, 2001.

Multiple Intelligences

Gardner, Howard. *Frames of Mind: The Theory of Multiple Intelligences.* New York, NY: Basic Books, 1993.

Gardner, Howard. *Intelligence Reframed: Multiple Intelligences for the 21st Century.* New York, NY: Basic Books, 1999.

National Science Standards

National Research Council. *National Science Education Standards.* Washington, D.C.: National Academy Press, 1995.

Share Your Bright Ideas

We want to hear from you!

Your name_____Date_____

School name_____

School address_____

City _____State_____Zip_____Phone number (_____)_____

Grade level(s) taught_____Subject area(s) taught_____

Where did you purchase this publication?_____

In what month do you purchase a majority of your supplements?_____

What moneys were used to purchase this product?

___School supplemental budget   ___Federal/state funding   ___Personal

Please "grade" this Walch publication in the following areas:

Quality of service you received when purchasing	A	B	C	D
Ease of use	A	B	C	D
Quality of content	A	B	C	D
Page layout	A	B	C	D
Organization of material	A	B	C	D
Suitability for grade level	A	B	C	D
Instructional value	A	B	C	D

COMMENTS:_____
_____

What specific supplemental materials would help you meet your current—or future—instructional needs?
_____

Have you used other Walch publications? If so, which ones?_____

May we use your comments in upcoming communications?   ___Yes   ___No

Please **FAX** this completed form to **888-991-5755**, or mail it to

Customer Service, Walch Publishing, P. O. Box 658, Portland, ME 04104-0658

We will send you a **FREE GIFT** in appreciation of your feedback. **THANK YOU!**